Florida's Water

Florida's Water:
A Fragile Resource
in a Vulnerable State

Tom Swihart

RFF PRESS
RESOURCES FOR THE FUTURE

New York ★ London

First published 2011 by RFF Press
2 Park Square, Milton Park, Abingdon, Oxon OX14 4RN

Simultaneously published in the USA and Canada
by RFF Press
711 Third Avenue, New York, NY 10017

RFF Press is an imprint of Taylor & Francis Group, an informa business

British Library Cataloguing in Publication Data
A catalogue record for this book is available from the British Library

Library of Congress Cataloging-in-Publication Data
Swihart, Thomas L., 1929-
    Florida's water : a fragile resource in a vulnerable state / Tom Swihart.
        p. cm.
    Includes bibliographical references and index.
    ISBN 978-1-61726-093-3 (hardback : alk. paper) 1. Water–Florida. 2. Water-supply–Florida. 3. Water quality–Florida. 4. Hydrologic cycle–Florida. 5. Water–Pollution–Florida. 6. Florida–Environmental conditions. I. Title.
    GB705.F5S94 2011
    363.6'109759–dc22

                                    2010045583

ISBN: 978-1-61726-093-3

Copyedited by Anda Divine
Typeset by OKS Press Services
Cover design by Chris Phillips
Printed and bound in the UK by Antony Rowe

The paper used is FSC certified.

About Resources for the Future *and* RFF Press

Resources for the Future (RFF) improves environmental and natural resource policymaking worldwide through independent social science research of the highest caliber. Founded in 1952, RFF pioneered the application of economics as a tool for developing more effective policy about the use and conservation of natural resources. Its scholars continue to employ social science methods to analyze critical issues concerning pollution control, energy policy, land and water use, hazardous waste, climate change, biodiversity, and the environmental challenges of developing countries.

RFF Press supports the mission of RFF by publishing book-length works that present a broad range of approaches to the study of natural resources and the environment. Its authors and editors include RFF staff, researchers from the larger academic and policy communities, and journalists. Audiences for publications by RFF Press include all of the participants in the policymaking process—scholars, the media, advocacy groups, NGOs, professionals in business and government, and the public. RFF Press is an imprint of **Earthscan,** a global publisher of books and journals about the environment and sustainable development.

About the Author

Tom Swihart has a master's in Urban and Regional Planning from Florida State University. His three-decade career in water management in the Florida Department of Environmental Protection included positions overseeing the adoption of state water quality standards, the department's water quality monitoring program, and the agency's Office of Water Policy. His duties as Administrator of the Office of Water Policy included updates of the Florida Water Plan, annual reports on the status of regional water supply planning across the state, revisions to the state water policy rule guiding the five regional water management districts, oversight on the adoption of minimum flows and levels for the state's diverse water resources, and the management of the state water conservation program for public water supply. He is a founding board member of the national Alliance for Water Efficiency.

Contents

Preface

*T*he Florida water management system has many unusual features and is fundamentally the same as when it was created almost four decades ago. Today, it confronts environmental and political challenges to long-term water resource sustainability hardly imagined in 1972. A water management system designed to meet the needs of 7.4 million Floridians needs updating to meet the demands in 2010 of almost 19 million. Fragile and vulnerable water resources, growing competition for finite supplies of water, serious water pollution problems, increasing flood hazards, a creaking property tax system for financing water management, and looming threats from climate change are all factors that prompt a need for changes. Of these threats, climate change is the supreme test because Florida is more vulnerable to its consequences than almost any other state in the United States. In my view, a water management system created in 1972, and designed to meet the challenges of that era, must be updated. This book explains how to do so without impairing the current strengths of the Florida water management system.

My perspective comes from three decades of service with the Florida Department of Environmental Protection, the state's primary water resource agency. I wrote this book because I know and care deeply about Florida. I have paddled, hiked, and camped in and around the state's extraordinary number and range of water resources. I have learned how vulnerable are the state's waters, and how much they deserve protection for future generations. My experience allows a unique look into those problems because no other book on Florida natural resource management – and very few books nationally – are written from such a perspective. I hope that the lessons of Florida water management will be meaningful for other states.

As Administrator of the Office of Water Policy, I gained a unique vantage point for learning how the state's five separate water management districts operate. I oversaw the preparation of the Florida Water Plan, as well as periodic updates to the state's "Water Policy" rule. I wrote guidelines for 20-year regional water supply plans for Florida, oversaw the distribution of funds to the water management districts for the development of alternative water supplies, and prepared annual reports on how well the state is doing in developing water supplies. I also co-authored the Department's *Framework for Action: Water Management and Climate Change in Florida*.

I also was the state's leader in creating "Conserve Florida," the state's water conservation program for urban water supply (carried out in partnership with the water management districts, water utilities, and other water conservation stakeholders). Through this, I learned how much water is wasted in Florida, how large is the potential for improvement, and how improved efficiency can forestall the need for expensive and wasteful water supply projects. I learned even more about that potential from serving on the founding board of the national Alliance for Water Efficiency, the country's premiere water conservation organization.

Water management looks different from the inside, in regard to both its successes and its failures. This book is a personal project and does not speak for the Florida Department of Environmental Protection. It does reflect, however, on what I learned in my profession of water management. One lesson is that water management is inherently value-laden rather than morally neutral and technocratic; getting the water "right" involves ethical choices. Another key lesson is that water management is a collaborative activity. Finally, although I recommend significant reform of some Florida water policies, that does not diminish the high regard I have for the current water management system or for the dedicated and competent professionals of the Florida Department of Environmental Protection and the regional water management districts.

Acknowledgments

This book would not have been possible without the support and assistance of many friends and colleagues. I feel enormous gratitude to the many people in the Florida Department of Environmental Protection, the water management districts, local government, nonprofit organizations, and the private sector who taught me about water management in Florida and who helped form this book.

Many people reviewed chapters or provided other help or inspiration. In that list, I must especially thank Carole Baker, Maribel Balbin, Cynthia Barnett, Bart Bibler, Jessica Bolson, Dave Bracciano, Lynn Coultas, Dana Bryan, Jack Davis, Mary Ann Dickinson, David Gilpin-Hudson, Wendy Graham, Kathleen Greenwood, Bevan Griffiths-Sattenspiel, Richard Haman, Dorota Hamann, Richard Harvey, Jim Heaney, Grace Johns, Pierce Jones, Steve Leitman, Helen Light, Janet Llewellyn, Rich Marella, Christopher Meindl, Kathryn Mennella, John Outland, Linda Pulliam, Kristen Riley, Dick Rubino, J. B. Ruhl, Katrina Schwartz, Steve Seibert, Jim Stevenson, Bruce Stiftel, Jessica Swihart, David Trimble, Clark Turner, Jake Varn, Amy Vickers, Sara Warner, and Kirk Webster. In his job as editor, Don Reisman shaped and improved the book in many ways. Dana Bryan did special service by reviewing almost every chapter and giving valuable comments each time. Chris Meindl also painstakingly reviewed the entire draft.

Lastly, this book would never have been written without the unwavering love, inspiration, and support of my wife and best friend Susan. She encouraged me to begin, contributed many key ideas, and read countless drafts. Discussing, challenging, editing, rewriting, encouraging, inspiring, and pushing, she has done as much as I have to get this book written. This book must be dedicated to her.

Note About Common References

OHP: Interviews from the Water Management, Growth Management, and Everglades Collections of the Samuel Proctor Oral History Program in the Department of History of the University of Florida.

Model Water Code: Frank E. Maloney, Richard C. Ausness, and J. Scott Morris, *A Model Water Code, With Commentary*, University of Florida Press, 1972. This book was the foundation used by the 1972 Florida Legislature in writing the Florida Water Resources Act.

Marella: *Rich Marella, Water Withdrawals, Use, and Trends Use in Florida*, 2005, U.S. Geological Survey, 2009.

Blake: Nelson Blake, *Land into Water — Water into Land*, University Press of Florida, 1980. This indispensable book was out of print for a number of years but the University of Florida reissued it in 2010 with updates by Christopher Meindl, Steven Noll, and David Tegeder.

St. Johns River Water
Management District

Northwest Florida
Water Management District

Suwannee River
Water Management District

Southwest Florida
Water Management District

South Florida
Water Mangement
District

Florida's five water management districts

Introduction

S hortly after the Suwannee River Water Management District was created in 1972, the new Governing Board of the north Florida district made a field trip 300 miles south to learn how the state's oldest and largest water management district approached water problems. The South Florida Water Management District and its predecessor agencies had already built or enlarged 1,400 miles of canals, 125 major spillways and dams, maintained the 100 mile-long Herbert Hoover Dike around Lake Okeechobee, and had recently converted the meandering Kissimmee River into a ruler-straight flood control channel 300 feet wide and 30 feet deep. They did this work with a staff of 900.[1] The Suwannee District had responsibility for no structures of any sort, and was a brand new organization. Nonetheless, it was responsible for meeting the water needs of people scattered across a large watershed. Its headquarter offices fit easily into the rooms of a small converted motel on the banks of the Suwannee River in the small town of White Springs. Offices included the convenience of a full bathroom.

Don Morgan, the Suwannee District's new executive director, led the tour because he formerly was the director of the Department of Planning in the South Florida District. The staff of the South District had conducted many such tours for other groups and were glad to assist in the educational process of inspecting canals, levees, and pump stations. An instructive event occurred when the Suwannee Board visited one of the giant flood control pump stations operated by the South Florida District. Combined, the large South Florida pump stations are capable of moving several billion gallons of water a day, comparable to the average flow of the Suwannee River itself.

One of the Governing Board members from north Florida asked what was the diesel fuel bill down south. The fuel cost for the South Florida Water Management District turned out to be more than the entire annual budget of the Suwannee River Water Management District. The visitors from north Florida were not favorably impressed with either the highly altered system in south Florida or the costs of operation. John Finlayson, a Governing Board member from the Suwannee District, is credited with creating a slogan that crisply expressed the alternative approach they would follow: *God made it and gravity runs it.*[2] The Suwannee District would not seek to rearrange its watershed by building and maintaining expensive dams, canals, and pump stations. Today, there is still no dam, dike, levee, or pump station on the Suwannee River in Florida.

This book tells the story of the Florida water management system that serves regions and water resources as different as the Suwannee River District with 310,000 residents and the South Florida District with 8 million. It asks whether that system can meet the population, water supply, water quality, and other challenges of a new century, including the unprecedented challenges posed by climate change.

In my view, the Florida water management system is reaching its limits. It is not yet prepared to meet the growing challenges of global climate change and other threats to sustainability. The product of almost 40 years of accretion since its creation in 1972, the present system has evolved as much as it can without major modifications. New ideas are needed to meet novel challenges in a state with extraordinarily valuable, yet fragile, water resources.

Part I: Framework, presents the overall structure of water management in Florida — how it got to where it is today.

- *Chapter One: The Watery Foundation,* describes the extraordinary scope, vulnerability, and diversity of the state's water resources. If Florida had been a desert rather than humid, rocky rather than sandy, or mountainous rather than flat, the challenges of water management would be quite different. The structure of water management that developed in recent decades is as much a response to hydrologic realities as were the first efforts in the mid-nineteenth century to drain the Everglades. The ideas held by Floridians about water management have been driven by the facts of water, filtered through unconscious assumptions and beliefs, and then reflected in subsequent political decisions.
- *Chapter Two: Origins,* explains how the current water management system was created in 1972. It took a remarkable confluence of circumstances to change the state's basic water law that year. The history of that event may have lessons for how to achieve equally bold changes today.
- *Chapter Three: The Responsibilities,* sets out in detail how Florida's water management system functions today. This book focuses primarily on the state's water management districts because they have very broad responsibilities, are unique in America, and their story has not yet been told at length. (To aid in understanding the similarities and differences between the water management districts, Appendix 1 provides a detailed chronology of water management milestone events.)

Part II: Celebrating Diversity, describes the different "personalities," problems, and achievements of the five water management districts. Each district faces challenges shaped by its region's particular water resources and its self-imposed habits of water management.

- *Chapter Four: Five Philosophies for Five Water Management Districts?* The five districts operate under the same statute but approach water management in distinctly different ways.
- *Chapter Five: God Made It and Gravity Runs It,* (Suwannee River Water Management District). Florida's most rural district is facing intensifying water quality and water supply challenges.
- *Chapter Six: The Forgotten District,* (Northwest Florida Water Management District). The interstate struggle on the Apalachicola-Chattahoochee-Flint River system is a primary focus of controversy for this water management district.
- *Chapter Seven: A Truce in the "Water Wars,"* (Southwest Florida Water Management District). "Swiftmud" is the locus of decades of water supply controversies in a major metropolitan region.
- *Chapter Eight: Reaching the Limits of Water Supply?* (St. Johns River Water Management District). This district is centered on the large river that defines the region and is now facing unprecedented water supply battles.
- *Chapter Nine: Managing Nature?* (South Florida Water Management District). The oldest and largest water management district includes the Florida Everglades, one of the nation's most endangered ecosystems.

Part III: The Issues, addresses the major issues confronting Florida water management in the twenty-first century, together with some recommendations on how to meet those challenges.

- *Chapter Ten: Water Supply Is Easy,* highlights the common but mostly unjustified fear that Floridians might be "running out of water." Popular misunderstanding about water supply fosters too many large and wasteful water supply projects. A real effort at water conservation, in addition to new policies to build drought resistance, would make the water supply problem in Florida much easier to solve.
- *Chapter Eleven: Saving the Water,* explains how improvements in water use efficiency have enormous potential for solving water supply problems – while also improving sustainability. It is the primary way in which water supply can be made "easy."
- *Chapter Twelve: Charge a Few Nickels for Water,* explains how water markets would not work for Florida and also how a water use fee system would put the water management system on a better financial and economic basis than would property taxes.
- *Chapter Thirteen: Farming the Water,* shows how Florida water policy provides too many incentives for agriculture to pollute and waste water.
- *Chapter Fourteen: Turning Up the Thermostat in Paradise,* describes how global climate change will dramatically affect Florida's water resources. Florida is at the center of a global atmospheric experiment gone wrong. Water management, energy, and climate change are closely interrelated. The state neglects those facts at its own peril.

Part IV: The Water Future, has a single chapter, *Chapter Fifteen: Will Florida Get the Water Future Right?* I hope that this book makes a contribution to making that possible.

NOTES

1. South Florida Water Management District. (1977) *Presentation to the House Natural Resources Committee,* February 8, 1977, p4

2. Oral History Program. (2005) John Finlayson, 7 July

PART I

THE FRAMEWORK
OF WATER MANAGEMENT

CHAPTER ONE

The Watery Foundation

*A*ll places depend on water, but no other place has Florida's particular combination of water resources, which has both enabled and constrained the state's pattern of human settlement. Its watery foundation is different from that of any other state. For sustainable water management, Floridians must understand the basic facts of the state's hydrology.

A VERY LONG PENINSULA

Jutting more than 400 miles into the sea, the Florida peninsula separates the Gulf of Mexico from the Atlantic Ocean. That feature of geography has profoundly shaped the state's water resources. There are 8,400 miles of shoreline (including bays and inlets), a total second only to Alaska. This lengthy coastline provides sandy beaches for tourists and helps define the state in popular imagination. The long coast also allows productive estuaries that foster a rich diversity of sea life. However, the sea is also very close to many wellfields barely above sea level. The sea is always threatening to turn wellfields salty if excessive fresh water withdrawals reverse the natural flow of groundwater and force saltwater to move inland.

The great length of Florida from north to south, with an ocean on one side and an enormous gulf on the other, results in multiple climate patterns. The humid subtropical northern section tends to have peak rainfall a month earlier in the summer than the southern section. The southern section also has a more pronounced dry season than northern Florida, which affects the availability of water supplies. Rivers to the northwest of a line from Jacksonville to Cedar Key tend to have their highest flow rates in late winter and spring, whereas rivers south of there tend to have their annual maximum flow in September and October.[1] Global warming will move Florida climate zones northward, producing a new tropical climate in much of the state. The full effects of this climate change cannot be predicted but certainly will mean many more water management challenges.

The peninsula configuration of the state shapes both the water resources of Florida and the human responses to them.

SHAPED BY THE SEA

The sea exerts a special power in Florida. No place in the state is more than 60 miles from the ocean. (And unless climate change is halted reasonably soon, the distance will be even shorter.) The place now called Florida is a fraction of a much larger geologic formation called the Florida Plateau, part of which is now above water; even more is below sea level today. For millions of years it has alternated between being above and below water. As the sea rose and fell, calcium and magnesium carbonate from oceanic organisms fell to the sea bottom along with the remains of countless generations of corals and mollusks. As geologic age followed geologic age, such material raining down grew deeper and deeper, forming characteristic limestone rocks rich in calcium. When sea levels fell, the limestone became the foundation of terrestrial Florida.

Ancient limestones now underlie surface soils but they are still interacting with water. These rocks are relatively soluble in water with acidity even as slight as that in natural rainwater. Rain percolating through Florida's organic soils can become even more acidic, further tending to dissolve the carbonate rocks. By geologic standards, this dissolution process is very rapid. Some groundwater basins in Florida dissolve limestone at rates of 60 to 140 tons of rock per square mile per year. At the higher end, the Silver Springs watershed system dissolves and removes 288 tons per square mile per year. This amounts to the dissolution and removal of an entire foot of limestone in a blink of geologic time: only 10,000 years.[2]

The dissolution of the underlying limestone in what is called a "karst" region creates both tiny pores and giant cavities. The voids in the limestone are usually filled with water and create some of the world's largest aquifers. The karst system also helps to create Florida's characteristic springs, lakes, and sinkholes when solution channels in the rock form or when giant cavities collapse.

WATER RESOURCES IN FLATLAND

Much of the geologic Florida Plateau has very little topographic relief, which directly affects water management. Florida's highest point on Britton Hill in Walton County is only 345 feet above sea level (the lowest maximum elevation of any state). Without much topographic relief, rivers tend to flow slowly. The "River of Grass," as Marjory Stoneman Douglas named the Everglades, is well known for having an extraordinarily small slope.

The gentle gradient of most Florida rivers has consequences for water management. In most other places in the United States, surface water reservoirs can be built at constrictions in river valleys. Valleys provide convenient and inexpensive sidewalls to help store water in reservoirs. In contrast, a place as flat as Florida has only limited opportunities to build conventional reservoirs. The nine existing river impoundments are shallow and tend to run out of water in droughts. To overcome these topographic disadvantages, water suppliers in Florida have constructed a number of large and expensive free-standing reservoirs adjacent to rivers to store water. However, all surface water reservoirs (in-stream or off-line) face engineering and water supply disadvantages in a hot state with high rates of evaporation.

A flat topography also matters for climate change. It means that rising sea levels in Florida will advance farther inland than in places with more steeply rising elevations next to the ocean.

IT RAINS A LOT

Florida has more thunderstorms than any other state. Much of the state has more than 80 days a year of thunderstorms.[3] Florida's average annual 54 inches of rain is only one inch less than Louisiana, the nation's wettest state. Among all states, Florida has the most rainfall in the month that is normally the wettest, as well as the highest maximum in a 30-minute rainfall.[4]

The state also experiences other phenomenal rainfall events. Over a single calendar year, 5 Florida weather stations have received 100 inches or more of rain. In one record month, 42 inches of rain fell in October, 1965 in Ft. Lauderdale.[5] Over shorter periods than years or months, deluges also occur. Rain events of more than three inches at a time account for an average of about 10 percent of total precipitation. In some northwest locations like Apalachicola and Pensacola, the annual average made by these high rainfall events is up to 17 percent of a year's rain.

High rainfall averages and intense rainfall events create many floods.

FLOODS

The urge to get rid of water is a recurring historical theme for Florida. Floods in a flat landscape, covering saturated soils, have nowhere to drain and can inundate regions not only for days, but also for weeks or months. The threat of flood damage in the state is extremely widespread even today. Floridians hold a highly disproportionate 41 percent of the entire nation's flood insurance policies, comprising $454 billion of insurance coverage.[6]

The potential for flooding is present throughout the peninsula. Among the defining water events in Florida's history were the disastrous hurricanes of 1926 and 1928, which caused many deaths on the shore of Lake Okeechobee. There had been warnings about flood hazards even before the 1926 flood deaths, but they were ignored. Only two years later, a hurricane caused Lake Okeechobee to overtop its crude flood dike. The death count from a failure to evacuate in time was more than 2,000 people. A congresswoman on an inspection tour afterward was horrified:

> Houses were folded together like a pack of cards. Some places you could see a foundation protrude without walls, here and there were houses upside down..., The people were still hunting for bodies. As many as had been recovered in the early days, many hundreds of them, were sent to the coast.... It soon became evident that they would have to burn the bodies as they were found, and they were piled in piles and covered with oil and burned.[7]

This event had national implications because it led to a new flood control role for the U.S. Army Corps of Engineers (USACE). In Florida, flooding has often been the crisis that changed water institutions, as is illustrated later in the origins of the state's five water management districts.

DROUGHTS

Many "dry" periods in Florida are "wet" by the standards of many other states. Florida droughts have, nonetheless, shaped the evolution of the state's water management system. Dry periods seldom occur at one time throughout the entire state. For more than a century, widespread drought has occurred at least once a decade somewhere in Florida.[8] The major drought of 1971 in south Florida was critically important to the creation of the modern water management districts. The drought of 1980–1981 was called at the time the "worst drought in South Florida history."[9] Other severe droughts followed. The drought of 1999–2001 resulted in statewide initiatives to develop new water supplies as well as renewed emphasis on water conservation. The recent drought of 2007–2009 focused attention on the same issues. Lake Okeechobee set a new record low of 8.82 feet above sea level on July 5, 2007. By August of 2008, the lake had gone more than 500 days below 11 feet, twice as long as the previous low period ever recorded. That drought eased. However, the state entered 2011 with a worsening drought.

Development has exacerbated the problem of droughts in Florida:

> Over south Florida during the past 100 years, there has been a widespread conversion of natural vegetation classes to urban and agricultural land, and into grassy shrubland. These landscape changes over south Florida are likely to have altered the local weather patterns, with south Florida averaged summer rainfall perhaps as much as 11% less than what would have occurred if the landscape had been left undisturbed by man.[10]

In droughts, Floridians can reach a point at which they almost wish for hurricanes to come if they deliver enough water. Climate change is expected to mean more variability in rainfall for Florida. Unusual weather will become more common or more extreme. If that occurs, more severe droughts will challenge the state's human occupants as well as all of its natural systems.

EXTREME EVENTS HAPPEN

Florida bears the brunt of more hurricanes than any other state. Eleven of the 12 most damaging hurricanes in North America affected Florida. Hurricane Andrew in 1992 produced insurance claims totaling almost $16 billion. The 2004 hurricane season was more active in Florida than any year since systematic recording began in 1851. Hurricane Charley alone damaged property from Port Charlotte to Daytona Beach, inflicting $13 billion in damages.[11]

Some tropical storms do most of their damage with winds well over 100 miles per hour, smashing both dense mangrove forests and gated subdivisions. However, some hurricanes carry with them deluges of rainfall, and most of the storm damage is from flooding. The actual rainfall delivered by a hurricane bears little relationship to the strength of the winds in the storm.[12] Smaller and less powerful tropical storms can deliver more rain than powerful hurricanes. In droughts, even destructive hurricanes may bring vital supplies of water if they are wet storms that refill aquifers and bring vital fresh water to estuaries.

Wet hurricanes can release extraordinary amounts of rain and cause massive flooding. The 1947 and 1960 tropical storms changed water management in two separate regions of Florida. The 1947 flooding in south Florida resulted in the creation of the Central and Southern Florida Flood Control District (predecessor to the South Florida Water Management District), and the 1960 damage inflicted by Hurricane Donna resulted directly in the creation of the Southwest Florida Water Management District.

A hurricane like the 1926 storm that hit Miami could cause up to $500 billion in damage by 2020, in light of today's intense coastal development.[13] Rising sea levels will greatly increase the damage from hurricane storm surges powering their way ashore. In Miami-Dade County alone, a sea level rise of only 2.13 feet makes the financial risk from hurricane inundation rise from $1 billion today to more than $12 billion.[14]

SURE IS HOT IN THE SUBTROPICS

On average, Florida has the highest January temperature and highest annual temperature of any state.[15] The high temperatures are significant for both population and growth patterns, but they also have many important consequences for water management in Florida. Lake evaporation in Florida ranges from 45–46 inches of water a year in north Florida to 50–54 inches in south Florida; water stored in reservoirs will quickly disappear unless rainfall is sustained.

The high temperatures make water management more difficult. Irrigation needs are much higher during hot weather, placing more pressure on water resources due to withdrawals. High temperatures accelerate biological activity, which depletes the oxygen needed for healthy rivers and lakes. Hot days also increase the rates of evapotranspiration, which means there is even less water available in streams and rivers to absorb pollutants.

FLORIDA GROUNDWATER

The aquifers of Florida are unusually widespread and productive. From a water supply perspective, groundwater can be even more valuable and much cheaper to harvest than surface water. The amount of water stored in an aquifer tends to fluctuate less than flow in a river. For the most part, groundwater has fewer biological contaminants than surface water. Water quality in an aquifer is relatively uniform, which simplifies the design and operation of drinking water treatment systems.

Water users in Florida recognize these advantages and have exploited the presence of immense groundwater resources. For example, a higher percentage of public water supply is derived from groundwater in Florida than any other state except Hawaii.[16] Florida is more dependent on groundwater than any other state east of the Mississippi River. The state withdraws more groundwater per square mile than all but three other states (Arkansas, Nebraska, and California). If the Southwest Florida Water Management District were a state, it would have a higher pumpage of groundwater per square mile than any other state.

Until recently, it was easy to turn to groundwater for water supply needs in most of Florida because usable quantities of groundwater can be found almost everywhere in the state. The 100,000 square miles of the Floridan Aquifer alone, extending into portions of Georgia, Alabama, and South Carolina, hold more than a quadrillion gallons of water. The enormous quantity in the Floridan Aquifer is about one-fifth as much as in all of the Great Lakes and 100 times that in Lake Mead on the Colorado River.[17] There is probably more water in Florida's aquifers than in any other state.[18]

Recharge to confined aquifers like the Floridan is not uniform across the state. Areas of high groundwater recharge encompass only one-sixth of the state.[19] This necessitates a land and water management system to identify and protect unusual recharge regions that benefit all of Florida.

SURFACE WATER IS GROUNDWATER IS SURFACE WATER IS GROUNDWATER

Unlike many other places, the boundary in Florida between surface water and groundwater is often indistinct. For example, there are a number of rivers that disappear by draining into sinks or "swallets" only to re-emerge several miles downstream. Even springs in Florida can reverse their flow in dry periods. When a large well begins to operate, it can lower water tables for miles around. If not managed properly, such wellfields can dry up nearby lakes and wetlands. The surface–groundwater connection is also shown in sinkholes, which puncture the impermeable layers above an aquifer and may transport surface water pollutants directly into the aquifer.

The close connection between surface water and groundwater is equally important for materials dissolved in the surface water. Fertilizers applied to the land surface often end up in aquifers, which then feed springs. The source of nitrogen varies among springsheds but agricultural and residential fertilizer applications are the principal sources statewide.[20] When the nutrient-laden water leaves the mouth of the spring and joins a river, dense algal mats and other excessive plant growth can grow in the spring run and the downstream river. Most Florida springs have already experienced this problem, which will persist for decades due to the slow movement of groundwater. Once loaded with nutrient pollution, groundwater is hard to clean up.

The long Florida coastline means that thousands of square miles of aquifers are at risk of saltwater intrusion. Florida is a hydrologic island of fresh water floating atop a sea of saltwater. If too much fresh water is pumped near the shore, the sea can intrude through the same porous limestone and sand that feeds the freshwater aquifer. The normal flow of groundwater flowing laterally toward the sea is what prevents saltwater from moving inland.

Lake Okeechobee is another prime example of this interconnection and interdependency between surface water and groundwater. This largest lake in Florida serves as a primary source of surface water for hundreds of thousands of acres of agricultural irrigation in the Everglades Agricultural Area (which is larger than the state of Rhode Island). Water from the lake is also a principal recharge source for the Biscayne aquifer, which is used by municipal wellfields for 6 million people in southeast Florida. These two expressions of Okeechobee as both surface water and groundwater illustrate the artificiality in Florida of separating the two—and how surface and groundwater must be managed as one system.

SPRINGS

Superlatives abound: there are more large springs in Florida than anywhere else in the United States or even the planet.[21] Even though many springs, of all sizes, have long been known in Florida, scientific study of them was sparse until recent years. In 1977, the total number inventoried was about 300 but sustained efforts in the last decade to identify springs have raised the total to more than 700. Thirty-nine of Florida's 67 counties have springs or have areas that contribute flow to springs.[22] There are 16 state parks named for springs. Ginnie Springs in Alachua County is the most popular freshwater diving site in the world. "First magnitude" springs, the largest in the classification system, have flows of greater than 100 cubic feet per second (equivalent to 64 million gallons a day). Thirty-three such first magnitude springs exist in Florida, which is a large fraction of the 84 known in the whole world.[23]

Unfortunately, many of the springs have problems, including flow reductions and high concentrations of nutrients. Too often, these impairments are related to intensive land use in the springshed that produces flow to the springs. Nitrate pollution damages many springs. For example, the Florida Department of Environmental Protection studied the water quality of 18 major springs in October 2000 through 2006.[24] All of the springs were on state or federal property. Nitrate/nitrite levels in the water were consistently high. That plant nutrient contributes to excessive plant growth in many Florida springs, especially algae.

Only four springs of the 18 appeared to have natural or near-natural levels of this nutrient. Many springs have nitrate levels high enough to be potentially toxic to some aquatic organisms.[25]

Misuse of land and inadequate water quality regulation are overwhelming the springs that Floridians value so highly and that help define the state.

HYDRO-BIOLOGY

The biota of Florida are exceptional, at least in part due to the unusual character and diversity of water resources. The wetland communities of the Everglades, for example, have developed in conditions of extremely low levels of available phosphorus. The ecosystems of Florida can tolerate or benefit from the state's cycles of flood and drought. For example, the seeds of the cypress trees that border many Florida lakes cannot germinate in standing water. They rely upon an occasional lengthy drawdown which exposes the dry lake bottom. When fully grown, however, cypress trees tolerate flooding better than any other tree in Florida.

Perhaps the most famous example of the ecosystem/hydrology connection is the Everglades, once home to astronomical numbers of wading birds. The birds were attracted by foraging conditions directly related to changing water levels. During the dry season their prey species were easier to catch in the shrunken pools of water. Severe droughts also killed many of the larger fish, allowing the smaller

Figure 1.1 *The survival of Florida wading birds like Snowy Egrets and White Ibises depends on specific cycles of water depth and availability*

fish favored by herons, egrets, storks, and other wading birds to become very numerous.[26]

Sudden flooding can have the opposite and damaging effect on bird nesting. The very high rainfall amounts in south Florida in May of 2009 harmed the wading bird populations during the critical part of the nesting season by making much of their foraging area inaccessible. This recovered very well in 2009 to near record levels and then fell back in 2010.The paradise for birds is dependent on particular seasonal, annual, and multi-year cycles of rainfall.

Climate change is likely to change those rainfall patterns significantly, with likely harmful effects on nesting birds and other species in Florida.

WET LAND MEANS WETLANDS

When North America was first settled by Europeans, wetlands were regarded as obstacles, hazards, and nuisances to be eliminated wherever possible. More than half of the wetlands in the lower 48 states were drained. Farming activities in the United States eliminated 42 million acres of wetlands.[27] The swamps and marshes in Florida were long regarded as wasteland, suitable only for conversion to other land uses. In 1850, Congress deeded to Florida more than 20 million acres of public land to "reclaim the swamp and overflowed lands" which were "unfit for cultivation."[28]

Before the era of wetland destruction began, well over half of the state of Florida was wetlands. By the end of the twentieth century, wetland acreage in Florida had been reduced from about 20.3 million acres to about 11.4 million, a loss of 44 percent. Even with all of the wetland losses, about 29 percent of the state is still wetlands, a higher percentage than any other state but Alaska.[29] Enormous wetland areas remain, such as the Everglades and the adjacent Big Cypress Swamp, the Green Swamp in the state's central plateau, the Big Bend coast on the Gulf of Mexico, the marshes of the upper and middle St. Johns River, and the floodplains of many other Florida rivers.

Boca Ciega Bay in Pinellas County illustrates the destructive trend as well as changes in Floridians' beliefs. A quarter of the Bay was filled or dredged by 1965. However, the filling and dredging occurred at a time when attitudes toward wetlands were changing. By the mid-1960s, wetlands were recognized as valuable resources and Floridians came to regret the damage done to Boca Ciega Bay. Shortly before Governor Leroy Collins died in 1991, he called the filling of the Bay a "monstrous desecration" committed over his opposition. On flights out of Tampa Bay, he always faced away from the coast so he would not have to see the harm done during his time in politics.[30]

In the 1960s "swamps" and "marshes" came more often to be called "wetlands," which sidestepped the older unfavorable connotations. By the 1970s, swamps once destined for draining had become wetlands worth preserving for future generations. Floridians today continue to struggle with what it means to live in a state so much of which is wetlands.

WETLANDS MEAN MOSQUITOES (AND DRAINAGE)

Some say that the state bird of Florida is the mosquito. Mosquitoes benefit from the same wet conditions that created rich organic soils and is so favorable for much of the state's agriculture. Urban dwellers have drained wetlands to reduce mosquito populations, just like farmers have drained wetlands to grow winter vegetables. Without mosquito control, much of Florida might be uninhabitable due to mosquito-borne illnesses. In 1922, Florida's chief medical officer put it starkly: "It is the plain and sober truth that we humans are engaged in a battle for self-preservation."[31] At the time, one of the most effective means of controlling mosquitoes involved shoveling dirt, because mosquito larvae need water to hatch. The solution was simple: Fill in wet areas or drain them. Mosquito populations went down after ditching but so did wetland acreage and estuarine productivity. Modern mosquito control programs are more sophisticated about the need to balance mosquito management with natural resource protection, but still are large operations. The Florida Keys Mosquito Control District, for example, has 71 employees. The Collier County District employs 26 staff members and operates a fleet of eight aircraft.[32]

Today, a new concern about mosquitoes and water is intensifying: climate change. The forthcoming effects on human health in Florida from climate change are unclear. However, the possibility of even more rainfall in a wet state may result in increasing mosquito numbers, along with the potential for more mosquito-borne diseases, as is described later.

SOILS

The water and soils of Florida are closely related to its development. Most of Florida is built on sand. It is for good reason that the most common lake name in Florida is "Sand" or "Sandy." Again, the basic hydrology is critically important: The hydrologic pattern helped to create the soils and humans took advantage of them. Florida has many regions of sandy soils which are low in nutrients, due in part to high annual rainfall dissolving and leaching materials out of the soil. There may be no other state with such extremes of regions that have either very low or very high amounts of available phosphorus, which is critical for plant growth.[33]

"Myakka fine sand" is the most common soil in the state and has been designated as the "official state soil" (Section 15.047, Florida Statutes). It is the soil propping up, for example, much of the state's citrus groves, which do poorly when their roots are immersed in water for too long. After the water table is lowered to make the soil more suitable for agriculture, the naturally sandy soils do not hold water long and dry out quickly when rainfall slows. Thus, irrigation is especially necessary in drained land—another of the paradoxes of Florida water management.

Some naturally wet soils of Florida, where the water table is frequently above ground surface, have high levels of organic matter formed naturally over the millennia. When drained, peat or muck soils have many advantages for agriculture. Yet after drainage, large areas of Florida's organic agricultural soils are lost due to subsidence: When the rich organic soils are drained and exposed to the air, they oxidize into the atmosphere. During the 5,000 years that the Everglades were forming, peat from dead plants accumulated at a rate of about 0.03 inches a year. In contrast, subsidence of farmed soils is estimated to occur at a rate of about 1 inch a year. Some land once 20 feet above sea level is now only 5 feet. That subsidence, caused by past and present drainage, complicates efforts to restore the Everglades.

In 1971, Art Marshall (one of the scientists most influential in calling for efforts to "repair" the Everglades, as he called it) viewed soil subsidence as a human-created catastrophe:[34] The chief engineer for the Central and Southern Flood Control District provided a perfect example of the contrast in views that has dominated the state's water management history. He took a much more utilitarian view in correspondence with Marshall. In his opinion, the hundreds of thousands of acres of organic soils of the Everglades Agricultural Area should be used in the same way as fossil fuels:

> You are correct that the muck is being consumed in those areas being farmed south of Lake Okeechobee. When used for these purposes the muck is a non-renewable resource. Like coal and oil, it is being used to satisfy man's energy requirements; in this case in the form of food.[35]

Today we have new concerns about soil oxidation and soil subsidence: greenhouse gas emissions. Carbon in soil is released to the atmosphere in the subsidence process. The wet soils of Florida still have the highest density of carbon in the continental United States and are close to the top in the mass of carbon stored in the soils.[36] It is possible that the trend of releasing carbon to the atmosphere from Florida soils could be reversed by conscious manipulation of hydrology and land use. Rather than being perceived as a longstanding obstacle to development, Florida wetlands could become an opportunity to make a unique contribution to slowing global climate change.

RIVERS

Close to 7 percent of American river miles are confined between constructed levees.[37] The lower 48 states have only 42 rivers that are more than 200 kilometers in length and are still free-flowing. Florida has four of these: the Choctawhatchee, Alapaha, Suwannee, and St. Marys, all of which originate outside of the state. None of these large Florida rivers has been designated as a National Wild and Scenic River. The two small rivers in the state that have been so designated are the Loxahatchee and Wekiva rivers, totaling only 49 miles of the 11,400 river miles in the National Wild and Scenic River system.[38] Florida failed to grasp the opportunities that this national program offers to protect the state's rivers.

Many of Florida's 1,700 streams and rivers, with a combined length of more than 26,000 miles, have been irreversibly altered.[39] Some rivers, like the

Hillsborough, Manatee, and Braden, have been impounded for urban water supply. Others have been channelized for flood control, like the Palm River in Tampa, which was completely destroyed when converted into the Tampa Bypass Canal. The Everglades drainage project in Dade County eliminated the small waterfall that once existed on the Miami River. More well-known is the project that turned the meandering Kissimmee River in the late 1960s into a straight and deep canal renamed by the USACE as C-38.

Although sandy beaches on the ocean are what attract the most tourists, the variety and number of rivers of Florida are more significant natural resources. The rivers are vulnerable to abuse because the flowing water in their channels is the summed result of land-use activities in their entire watershed. Only the management of entire watersheds can protect Florida's remarkable rivers, which are described individually in later chapters.

RIVERS MAKE ESTUARIES

Estuaries are where fresh water meets and mixes with the waters of the ocean, such as in salt marshes, bays, or river lagoons. They are dynamic creations, depending on alternating cycles of fresh and salty water to make one of the most productive ecosystems in nature. For example, all but 3 of Florida's 28 national wildlife refuges are on the coast. About 85 percent of Florida's recreational and commercial fish species spend part of their life cycle in estuaries. Their coastal seagrass beds are the nursery of many fish species as well as other important species like blue crabs and scallops. The organisms and ecosystems of estuaries are adapted to naturally variable flows. Unnatural surges (as from sudden releases from reservoirs upstream), or too little fresh water (as from impounding vital waters upstream) can damage the delicate balance of life in the estuaries. Estuaries provided almost all of the food eaten by the sixteenth century Calusa Indians of southwest Florida, one of the few tribes in the southeastern United Stated that did no farming. They turned down offers of agricultural tools from Spanish invaders, saying that they had no use for them.[40]

Despite the many alterations of Florida rivers over the last century, their flow still fosters some of the largest and most productive estuaries in the United States and the world. The seagrass expanses of the Indian River Lagoon, for example, provide at least $5,500 a year per acre in total economic value to the region.[41] Florida Bay, at the tip of the peninsula, is North America's largest subtropical estuary, totaling 850 square miles. Florida has 3 of the nation's 28 National Estuarine Research Reserves (Apalachicola, Guana-Tolomato-Matanzas, and Rookery Bay) with a total area of 643 square miles. In addition, 4 of the 28 National Estuary Programs are in Florida (Charlotte Harbor, Indian River Lagoon, Sarasota Bay, and Tampa Bay), which is more than in any other state.

Without adequate amounts of water – of the right quality – Florida estuaries become much less productive and valuable. If the rivers or aquifers supplying estuaries with fresh water are diverted for human uses, estuaries suffer. Rising sea levels will threaten both the estuaries and the entire downstream portions of the rivers that feed them.

RAIN MAKES LAKES

Thoreau called lakes the landscape's "most beautiful and expressive feature" in which the beholder "measures the depth of his own nature."[42] Nowhere else in the southern United States is there a lake district like Florida. There are 12 counties with more than 200 lakes. Only 19 counties have fewer than 30 and there is no county without at least one lake.[43]

Most lakes in Florida are relatively shallow basins formed by dissolution of the underlying limestones (often called "solution depressions"). Many are geometrically circular as a consequence of the sudden collapse of underground cavities. Only a small fraction of the array of lakes is more than a few dozen feet deep. Most lake basins are relatively flat; small fluctuations in lake levels result either in the exposure or the submergence of large expanses of lake bottom. This shrinkage and expansion of the littoral zone is actually very healthy for lakes, consolidating loose sediment and revitalizing lake ecosystems. The lakes are rejuvenated by the wet–dry cycle.

By Florida standards, Okeechobee really is the "big water," as the Miccosukee tribe of Native Americans called it. Lake Okeechobee is the state's largest and most famous lake. It has an average depth of only 9 feet and is only about 18 feet deep at flood stage. Because it is shallow, the surface area of the lake varies dramatically with small changes in depth. At 17 feet elevation above sea level in late 2005, the lake covered 712 square miles. Only 20 months later, when the lake had fallen to 8.82 feet, the lake covered only 463 square miles. Okeechobee, depending on the level of the lake, holds about one cubic mile of water. This sounds like a large volume until it is compared with the 2,935 cubic miles of Lake Superior. Okeechobee's volume is very small relative to some lakes of roughly comparable surface area because it is much shallower.

HYDRO-IDEOLOGY

The state has both unique water resources and a unique history of human manipulations of its water. Floridian beliefs and attitudes about their extraordinary water resources have led to some narrow-visioned decisions, including massive efforts to drain the Everglades. For example, the 1958 annual report of the Central and Southern Florida Flood Control District reflected the conventional view at the time that Florida was made to be exploited:

> Ever since white man first set eyes on the inland areas of the lower part of Florida, he has dreamed of bringing it under control for agricultural use. Its rich soils and the mild weather that prevails generally impressed him with its fantastic possibilities.[44]

However, a contrary view also runs through the narrative of water management in Florida — that the state is a natural paradise to be preserved. Former Governor and Senator Bob Graham expressed concern in 2011 that politicians may be resurrecting outmoded ideas about the watery foundation of Florida:

> For two-thirds of the 20th century, Florida was defined as a commodity. If you thought that it was too wet, you filled it in. If you thought it was too dry, you dug it up and packaged it and sold it. [In the 1960s environmentalists urged people to stop thinking of Florida as a commodity and start thinking of it] as a treasure ... That's the battle many of us thought was settled and which has now re-emerged. It is now one, if not the, central issue of Florida politics.[45]

NOTES

1. Florida Department of Environmental Protection, Bureau of Watershed Management. (2006) *Integrated Water Quality Assessment for Florida: 2006 305(b) Report and 303(d) List Update*, Florida Department of Environmental Protection, Bureau of Watershed Management, Tallahassee, FL, p7

2. Fernald, Edward A., and Donald J. Patton. (1984) *Water Resources Atlas of Florida*, Florida State University, Tallahassee, FL, p46

3. Fernald, Edward A., and Elizabeth D. Purdum. (1994) *Water Resources Atlas of Florida*, Florida State University, Tallahassee, FL, p26

4. Henry, James A., Kenneth M. Portier, and J. Coyne. (1994) *The Climate and Weather of Florida*, Pineapple Press, Sarasota, FL, pp101 and 123

5. Fernald and Purdum, op. cit. note 3, p20

6. Committee on Community Affairs, Florida Senate. (2008) *Agency Sunset Review of the Department of Community Affairs. Issue Brief 2009-310, October 2008*, Division of Legislative Information Services, The Florida Senate, Tallahassee, FL, p29

7. Congresswoman Ruth Bryan Owen, quoted in Huser, Thomas (1990), *Into the Fifth Decade: The First Forty Years of the South Florida Water Management District, 1949–1989*, South Florida Water Management District, West Palm Beach, FL, p1

8. Winsberg, Morton D. (1990) *The Weather of Florida*, University Press of Florida, Gainesville, FL, p83

9. Huser, Thomas. (1990) *Into the Fifth Decade: The First Forty Years of the South Florida Water Management District, 1949–1989*, South Florida Water Management District, West Palm Beach, FL, p115

10. Pielke, R. A., L. T. Steyaert, P. L. Vidale, G. E. Liston, W. A. Lyons, and T. N. Chase. (1999) "The influence of anthropogenic landscape changes on weather in south Florida," *Monthly Weather Review*, vol. 127, p1669

11. Florida Catastrophic Storm Risk Management Center, College of Business, Florida State University. (2010) "Florida's property insurance landscape," www.stormrisk.org/index.cfm?page=3, accessed 3 July 2010

12. Fernald and Purdum, op. cit. note 3, p28

13. Karl, Thomas R., Gerald A. Meehl, Christopher D. Miller, Susan J. Hassol, Anne M. Waple, and William L. Murray (eds). (2008) *Weather and Climate Extremes in a Changing*

Climate. Regions of Focus: North America, Hawaii, Caribbean, and U.S. Pacific Islands, A report by the U.S. Climate Change Science Program and the Subcommittee on Global Change Research, Washington, DC

14. Harrington, Julie, and Todd L. Walton, Jr. (2008) *Climate Change in Coastal Areas in Florida: Sea Level Rise Estimation and Economic Analysis to the Year 2080*, Florida State University, Tallahassee, FL, pp1–49

15. Henry et al, op. cit. note 4, p7

16. Kenny, J. F., N. L. Barber, S. S. Hutson, K. S. Linsey, J. K. Lovelace, and M. A. Maupin, (2009) *Estimated Use of Water in the United States in 2005: U.S. Geological Survey Circular 1344*, U.S. Geological Survey, Washington, DC

17. Heath, R. C., and C. S. Conover (1981) *Hydrologic Almanac of Florida*. U.S. Geological Survey Open-File Report 81-1107. Tallahassee, FL, p39

18. Fernald and Purdum, op. cit. note 3, p38

19. Purdum, Elizabeth D. (2002) *Florida Waters: A Water Resource Manual from Florida's Water Management Districts*. Communications and Community Affairs Department, Southwest Florida Water Management District, West Palm Beach, FL, p49

20. Cohen, Matthew J., Sanjay Lamsal, and Larry V. Kohrnak. (2007) *Sources, Transport and Transformations of Nitrate-N in the Florida Environment*. St. Johns River Water Management District, *Special Publication SJ2007*, St. Johns River Water Management District, Palatka, FL

21. Knight, Robert L., and Sky K. Notestein. (2008) "Florida's springs – What do we know?" *Florida Watershed Journal*, vol. 1, no. 2, pp11–15

22. Florida Department of Environmental Protection. (2008) *Florida Springs Initiative: Program Summary and Recommendations, 2007*, Florida Department of Environmental Protection, Tallahassee, FL, p3

23. Purdum, op. cit. note 19, p49

24. Florida Department of Environmental Protection. (2008) *Springs Long-Term Water Quality*, Florida Department of Environmental Protection, Tallahassee, FL

25. Mattson, Robert A., May Lehmensiek, and Edgar F. Lowe. (2007) *Nitrate Toxicity in Florida Springs and Spring-Run Streams, St. Johns River Water Management District, Professional Paper SJ2007-PP1*, St. Johns River Water Management District, Palatka, FL

26. Levin, Ted. (2004) *Liquid Land: A Journey through the Florida Everglades*, University of Georgia Press, Athens, GA, p165

27. McNeill, J. R. (2000) *Something New under the Sun: An Environmental History of the Twentieth-Century World*, W. W. Norton, New York, p187

28. Internal Improvement Fund of the State of Florida. (1980) "Minutes of the proceedings of the board of trustees of the Internal Improvement Fund of the State of Florida," in Nelson Blake, *Land into Water – Water into Land*, (1980) University Press of Florida, Gainesville, FL, p36

29. Dahl, Thomas E. (2005) *Florida's Wetlands: An Update on Status and Trends, 1985 to 1996*, U.S. Fish and Wildlife Service, Washington, DC

30. Stephenson, R. Bruce. (1997) *Visions of Eden: Environmentalism, Urban Planning, and City Building in St. Petersburg, Florida, 1900–1995*, Ohio State University Press, Columbus, OH, p142

31. Col. Raymond Turck, quoted in Gordon Patterson (2004) *The Mosquito Wars: A History of Mosquito Control in Florida*, University Press of Florida, Gainesville, FL, p33

32. Florida Keys Mosquito Control District (2010) www.keysmosquito.org/ and Collier County Mosquito Control District (2010) www.cmcd.org/, both accessed 14 July 2010

33. Myers, Ronald L., and John J. Ewel. (1990) *Ecosystems of Florida*, University Press of Florida, Gainesville, FL, p63

34. Arthur R. Marshall. (1971) "Repairing the Florida Everglades basin," *The Florida Naturalist,* June 11, p2

35. Storch, W. V., letter of July 6, 1971 to Arthur R. Marshall, replying to Marshall's June 11, 1971 statement in "Repairing the Florida Everglades basin," *The Florida Naturalist* [refer to note 34]

36. Curry, Susan. (ed). (2007) *Myakka*, Soil and Water Science Department, University of Florida, Gainesville, FL, p2

37. McNeill, op. cit. note 27, p183

38. U.S. Fish and Wildlife Service. (2010) "National Wild & Scenic Rivers System," www.rivers.gov/waterfacts.html, accessed 12 August 2010

39. Florida Department of Environmental Protection. (2010) *Draft Integrated Water Quality Assessment for Florida: 2010 305(b) Report and 303(d) List Update*, Florida Department of Environmental Protection, Tallahassee, FL

40. Marquardt, William H. (2004). "Calusa," in Raymond Fogelson and William Sturtevant (eds), *Handbook of North American Indians*, vol. 14, "Southeast." Smithsonian Institution, Washington, DC, p206

41. Johns, Grace. (2008) *Indian River Lagoon Economic Assessment and Analysis Update.* Final report for St. Johns River Water Management District and South Florida Water Management District, Indian River Lagoon National Estuary Program, Contract No. 24706, St. Johns River Water Management District, Palatka, FL

42. Thoreau, Henry David. *Walden,* chapter IX, "The ponds"

43. Heath, op. cit., note 17, p60

44. Central and Southern Florida (C&SF) Project. (1957) *Central and Southern Florida Flood Control Project, Eight Years of Progress, 1948–57 Report*, South Florida Water Management District (formerly the Central and Southern Florida Flood Control District), West Palm Beach, FL, pv

45. Klas, Mary Ellen "Former Sen. Bob Graham warns that oil spill danger hasn't passed", *St. Petersburg Times,* January 14, 2011

CHAPTER TWO

Origins of the Modern Water Management System

T he dream of remaking Florida's water resources began early. In 1821, the year the United States acquired Florida from Spain, Col. James Grant Forbes suggested that Lake Okeechobee could become the "grand central source of communication between the Atlantic and Mexican Sea."[1] When Florida entered the Union in 1845, it was a frontier state and economic development was a priority. Balanced at admission to the Union as a slave state against the free state of Iowa, Florida's 70,000 white citizens were keenly interested in profiting from both human property and the available natural resources. There was little scientific understanding of the state's water resources and a general belief that flood control and drainage were far more important than protection of what seemed to be an endless supply of nearly useless wetlands. Admission to the Union opened wide opportunities for converting Florida land and water resources into economic prosperity, including an immediate grant from the national government of 500,000 acres. That was only a tiny fraction of what was to come.

Floridians encouraged the federal government to give up most of its land holdings. That happened for Florida and other states in 1850 by the passage of the federal Swamp Lands Act. The national government transferred "swamp and overflowed lands" to the states for the purpose of draining them and making them more productive. Eventually, the allotment to Florida under this Act alone totaled 20.3 million acres. Other federal statutes drove the total of federal lands transferred to Florida to 24 million acres, more than that received by any western state.

After the devastating Civil War, the state still held many millions of acres. Northern businesspeople struck deals with the financially troubled state government. In one case, the state sold a million acres to a New York company for only 10 cents per acre. A single individual, Hamilton Disston, agreed in 1876 to drain 12 million acres in exchange for ownership of 6 million acres. In 1881, Disston and the state additionally agreed that he would buy 4 million acres for only

25 cents an acre. His Florida Land and Improvement Company opened sales offices in every major American city.[2]

Disston brought two enormous dredges to the state and placed one in Fort Myers at the mouth of the Caloosahatchee River and one in Kissimmee. The plan was to dredge a drainage path from the Kissimmee River basin to Lake Okeechobee and also from the big lake to the Gulf of Mexico. Disston's dredges eventually connected several lakes of the upper Kissimmee basin and straightened the meandering Caloosahatchee. His ambitious projects were cut off by the Panic of 1893, which dried up sales and prevented refinancing. The state government took back some of the land for taxes; the rest was sold for a pittance of what he had paid.

The state acquired the final major block of 2.8 million acres of federal "Swamp and Overflowed" land in 1901, mostly located in the Everglades. Efforts to subdue the Everglades extended beyond dredging. For example, John Gifford, the first American to earn a doctorate in forestry, was a strong advocate of using melaleuca trees (*Melaleuca quinquenervia*) to drain the Everglades. He thought that non-native trees like the melaleuca could evapotranspire so much water that they could dry up the region's wetlands. He called upon Floridians to conquer nature:

> In Southern Florida we have the resources, but the vim has been lacking. We have been reposing since the Seminole war. It is not laziness. We have been indulging our love of leisure. But it is this grappling with nature which develops the latent forces within the man. The coming age is to be an age of conquest, the conquest of nature, the reclamation of swamp lands and the irrigation of deserts.[3]

Today, melaleuca is a major invasive pest species rather than proof of human ability to conquer the natural world. The introduced tree did not drain the Everglades, but forced the expenditure of many millions of dollars in an attempt to control it. Although this early effort at waging biological war on wetlands failed, the state could still turn to mechanical dredges to accomplish the same purpose. In the early 1900s, the Florida legislature helped create scores of local "drainage districts" intended to "reclaim" land at minimal state expense. Trying to transform the Everglades on the cheap also led to the construction of a crude 5-foot-high levee around Lake Okeechobee. The new dike was built not of expensive rock or cement, but from the sand and mud next to the dike itself.

Other parts of Florida also were busy with water resource modifications. The Disston era included many navigation and drainage channels in the Apopka-Ocklawaha River region of central Florida. Lake Apopka, the third-largest lake in the state, served as a repository in the 1920s for citrus processing plant waste and raw sewage from the city of Winter Garden. In the early part of World War II, fears of a looming "food shortage" were used to justify the drainage of about 20,000 acres of the northern edge of Apopka. At the end of World War II, the 1945 State Committee on Water Resources advocated the creation of a Florida Department of Water Resources for "fact finding, planning, promoting and protective work, relating to fresh water resources and problems." The Committee believed that the "state must make an immediate start on a vigorous program to conserve, protect, develop, control, and utilize its water resources for the public welfare."[4] Although eventually the state created such a department, it gave the agency few powers or resources. As has often been the case in Florida, contemporary water events, such as massive flooding in 1947, overtook planning.

That year, summer storms flooded the rich farming areas of the Everglades. A hurricane in October intensified the agricultural damage but had even more serious impacts on developed regions of the entire southeast coast. Two feet of water covered the streets of Fort Lauderdale. More than 10,000 people had to be evacuated from their homes. Floridians realized, too late, that the defective and incomplete flood control system could not meet the needs of an urbanizing state. Politicians, newspapers, and other community leaders demanded new flood control plans. However, implementation would cost far more than a rural state could afford. The essential partner again, as in the building of a stronger dike around Lake Okeechobee in the 1930s, was the federal government, represented by the U.S. Army Corps of Engineers. State interests and the Corps agreed in 1948 on the scope of a new "Central and Southern Florida Project for Flood Control and Other Purposes" (C&SF Project) with three main components:

> First, it established a perimeter levee through the eastern portion of the Everglades, blocking sheet flow so that lands farther east would be protected from direct Everglades flooding. The levee was about 100 miles long. It became the westward limit of agricultural, residential, and other land development for the lower east coast from West Palm Beach to Homestead. Only a few areas were subsequently developed west of it, most notably the 8.5 square-mile residential area of Dade County south of the Tamiami Trail, northwest of Homestead. The perimeter levee severed the eastern 16 percent of the Everglades from the interior.

> Second, the C&SF Project designated a large portion of the northern Everglades, south of Lake Okeechobee, to be managed for agriculture. Only a portion near the lake had previously been developed, leaving much room for agricultural expansion in what was a vast, nearly unbroken expanse of sawgrass. [...] Called the Everglades Agricultural Area (EAA), it encompassed about 27 percent of the historic Everglades and was a major factor in the economic justification of the C&SF Project.

> Third, water conservation became the primary designated use for most of the remaining Everglades between the EAA and Everglades National Park, limited on the east by the eastern perimeter levee by an incomplete levee bordering the Big Cypress swamp. The area was divided into three units (essentially wetland impoundments) called Water Conservation Areas (WCAs).[5]

The Army Corps of Engineers designed the project but required that a local party take responsibility for sharing the costs and for operation and maintenance. The demand for flood control was so urgent that both Congress and the Florida legislature acted as early as the end of 1949 to authorize the Corps' project and a new regional agency called the Central and Southern Florida Flood Control District. Completion of the entire project was expected to take about a decade.

Figure 2.1 *In 1950, Floridians celebrated the start of construction on the Central and Southern Flood Control Project; in the 21st century, the South Florida Water Management District, the state of Florida, and the federal government are undertaking to redesign it*

Like previous proponents of Florida water projects, advocates of the C&SF Project had very high hopes for success.

In the mid-1950s, another state commission tried to figure out what to do about Florida's long-standing water problems. In 1956 the Water Resource Study Commission recommended that more attention be paid to statewide water management but saw only a few of its recommendations implemented when the legislature passed the Water Resources Act of 1957. The legislature took very small steps to prevent water users from taking too much water out of surface and groundwaters, and the limits were so lenient as to be almost meaningless. For example, water users were allowed to reduce aquifer levels to half the distance between sea level and the average groundwater elevation during the previous 20 years.[6]

Flood control and drainage were emphasized rather than watershed management. Just as the 1945 Committee on Water Resources was followed a few years later by regional flooding in 1947 in southeast Florida, implementation of the Water Resources Act of 1957 was followed in 1959 and 1960 by serious flooding in the Tampa Bay region. Almost 1 million acres were flooded, causing massive agricultural disruption and urban damage. Following the example of the

C&SF District, local interests, the state legislature, and the Army Corps of Engineers jointly designed a regional flood control system called the Four Rivers Basin Project. Flood detention systems and flood control channels were to be constructed on the Hillsborough, Oklawaha, Withlacoochee, and Peace Rivers. The 1961 legislature created the Southwest Florida Water Management District (SWFWMD) to take on the management responsibilities, in a manner akin to the Central and Southern Florida Flood Control District.

The SWFWMD was deliberately named a water *management* district, unlike the Central and Southern *Flood Control* district to signal that it was a multiple-purpose entity.[7] The old-fashioned structural approach to water management continued to dominate much of Florida, however. Natural resources were deemed to be of little value unless they were altered. Many still believed that even more dikes, ditches, and pump stations would tame the state's unruly water resources. As late as 1963, the state Board of Conservation recommended dozens of new impoundments and canals in all regions of the state.

The same political forces that tried to drain the Everglades in the nineteenth century for farming were unchallenged far into the twentieth century. According to Nathaniel Reed, former Assistant Secretary of the Interior and member of the Governing Board of the South Florida Water Management District, the Central and Southern Florida Flood Control District was "totally dominated" by agricultural and industrial interests.[8] On a broader scale, the Florida Legislature was a failed democratic institution. As Gene Burnett observed:

> They were the "Pork Chop Gang" – truly a mutation of representative government – for they represented more pine trees than people and, with control of less than 15% of the state's population, they could elect and control a majority of the state legislature. Florida's antiquated constitution[...] decreed that, regardless of population, each county could have no more than one senator and no more than three representatives.[9]

One state representative in north Florida had only 2,199 constituents, while another representative in south Florida had 165,028. People residing in Florida Senate districts had a similar spread: from as low as 10,413 in north Florida to as much as 495,084 in south Florida.[10] The Pork Choppers were conservatives from rural Northern and Central Florida, committed to "states' rights" and the archaic legislative district apportionment specified in the state's 1885 constitution. In his recent intensive study of their history, Seth Weitz said that Pork Choppers viewed themselves as "the last bastion of protection for the agrarian lifestyle of rural Florida which was being challenged by growing industry and big business. They played on the fears of Floridians, "creating the 'other', whether it was African-Americans or liberals or later communists and homosexuals, in an attempt to rally support behind their cause."[11] They were zealous proponents of keeping Florida a state with very low taxes and minimal regulation.

The U.S. Supreme Court's "one-man, one-vote" ruling in *Baker v. Carr* in 1962 led lower courts in the late 1960s to correct Florida's malapportionment. Until *Baker v. Carr*, Florida's legislature was the most malapportioned and unrepresentative in the country.[12] Reapportionment drastically reduced the numbers and power of the Pork Choppers by way of an unusually large number of first-time legislators and fresh ideas, including later Governors Reubin Askew and Bob Graham. Nathaniel Reed noted the electoral consequences:

> We had the shortest election period that has ever been known and attracted some of the best – both Republicans and Democrats – some of the brightest young men and women who ever served in Florida government came together for a minimum of four years. We had the most spectacular legislature that was almost untouched by the lobby[ists]. They were men and women who really brought new ideas to Tallahassee.[13]

In a political instant, politics in Florida changed. In 1972, the "Year of the Environment," Florida leapt into the national lead on state efforts at growth management. In that one year, the legislature reorganized the state's natural resource management agencies and also passed the Environmental Land and Water Management Act, the State Comprehensive Planning Act, and the Land Conservation Act. Equally significant, they created major reforms in water management in the Water Resources Act of 1972.

A major drought had troubled south Florida in 1971. From October 1970 until May 1971, the Miami airport had only 2.04 inches of rain. Saltwater intrusion closed eight Miami drinking water wells. The Central and Southern Florida Flood Control District asked users to cut back on their withdrawals but had little authority to require real reductions.[14] The entire country was in ferment over the Vietnam war, the rise of environmentalism (the first Earth Day was in 1970), civil rights, and women's rights. The major drought in south Florida began in 1970 against this background of human turmoil. The drought intensified in 1971 and Reuben Askew, the new Democratic governor, orchestrated an "American Assembly" meeting in Miami to devise solutions to the water problems.[15] John DeGrove was co-chair of the meeting with Art Marshall. DeGrove later explained, "What brought these people to the table was Askew making it clear that whatever this conference recommended, he was going to push very hard to get it passed in the upcoming 1972 session."[16]

The strong support from Governor Askew made the difference. A recent study of water policy change in Arizona observed that state water policies usually change only incrementally. Normally, only a few alternatives to the current policy environment are considered and only slight changes may be implemented. Only a strong governor can shake the grip of tradition.[17] Governor Askew provided that leadership at a time in Florida history that was open to other kinds of growth management changes.

The Assembly, unlike most such meetings, made a big impact. John DeGrove later said, somewhat optimistically, that it marked the "end of Florida's uncritical love affair with growth."[18] The 150 participants agreed to a wide-ranging and action-oriented set of findings and recommendations. The timing was perfect. The focus on water policy reform was bolstered by the remarkable coincidence that University of Florida law professors had just finished drafting a Model Water Code that was available for consideration as legislation.[19] (See the next chapter for more on the Model Water Code and how it is reflected in Florida water law.) Frank E. Maloney and Richard C. Ausness (University of Florida) and J. Scott Morris (Southern Methodist University) were the principal authors. Jacob D. Varn, a former staffer of the Southwest Florida Water Management District and a law student at the time, drafted the chapters on regulation of wells and surface water.[20]

Governor Askew did not merely accept the report from the conference. He immediately appointed a Task Force on Resource Management to prepare implementing legislation not only for water management but also for a host of related comprehensive and land use planning matters. Members included key participants in the American Assembly meeting, such as John DeGrove (who acted as chair), Art Marshall, and state senator (later governor) Bob Graham.[21] The Task Force set the stage for the 1972 legislative enactment of what Governor Graham later called a "quiet revolution" in growth management.[22]

The scholarly work under way at the University of Florida on a Model Water Code became the framework for fundamental reform of Florida's water law in the spring session of the legislature in 1972. Key individuals lobbied for enactment, like Hal Scott (executive director of the Florida Audubon Society), Marjorie Carr (executive director of the Florida Defenders of the Environment), and Betty Castor (president of the League of Women Voters, later state Education Commissioner and president of the University of South Florida). Despite broad support, the proposed reforms encountered strong opposition. The St. Joe Paper Company, the largest landowner in the state, was a prominent opponent, as were the electric utility companies. Agricultural interests also were uneasy or opposed, according to Derrill McAteer, a chair of the Governing Board of the Southwest Florida Water Management District:

> Agriculture [opposed it] because they felt they were going to lose the water. I think that we got agriculture finally to wake up. My main theory [in] every speech I gave was, gentlemen, this is your ticket, this is your license. For God's sake, don't turn your back on it because ultimately the cities are going to take it away from you. Get on this path. And I think the leadership of agriculture probably came around. I was called a traitor. I was called everything in the world.[23]

The water management districts were created in the political earthquake of the early 1970s, along with other new Florida institutions. Senator Phil Lewis, one of the key parties in the creation of the Water Resources Act of 1972, pointed to the districts' importance: "I'm starting to tell everyone interested in politics and power to watch the water management districts and their boards. That's where the power is going to be."[24] That would not become true unless the new districts had adequate funding to carry out their statutory tasks. Instead of the funding situation getting better, it got worse after 1972. The water management districts existed as legal entities, but there was hardly any money available for the three new ones: St. Johns River, Suwannee River, and Northwest Florida. The Model Water Code and the Florida statute did not solve the problem of how to provide adequate and stable funding for water management districts. The Code recommended a vague combination of ad valorem taxation, state appropriations, and permit "surveillance fees" as funding sources.

The ability to levy property taxes had not been a problem for the Central and Southern Florida Flood Control District (which became the South Florida Water Management District as a result of the new statute) and the Southwest Florida Water Management District, which were assuming new responsibilities; they already had the power to levy property taxes. A court found in 1975 that the statutory change of boundaries for the already-existing Central and Southern Florida Flood Control District and the Southwest Florida Water Management Districts voided the property taxation powers they had under their old boundaries.

The three new districts had no such power at all. The Florida legislature proved unwilling to fund their new districts at a level that would make them capable of performing the many tasks assigned them by the new statute. Without substantial funding available on a secure and recurring basis, the new districts were unable to function adequately or plan for long-term water projects. The legislature postponed full implementation until December 31, 1976. Don Feaster, the executive director of the Southwest Florida Water Management District at the time, credited Buddy Blain with the idea of how to fund the districts properly:

> I was in a meeting when we were discussing funding for the Water Management District, maybe in [19]73 or [197]4 in Tallahassee, a room full of people from all of the water management districts and the staff of, I believe, out of the Board of Conservation, before it was DNR, but in any case, there was discussion how do we get funding. Nobody had an idea. [Buddy Blain] proposed a constitutional amendment, and everybody in the room proceeded to say he was crazy. He laid out a strategy, and everybody ended up agreeing with that strategy. It ended up becoming a constitutional amendment that overwhelmingly passed, and without that, I do not believe that today the other three new water management districts would be anything but paper.[25]

This was a bold idea because it would require a majority vote on a constitutional amendment in a state where politicians usually promised to reduce, rather than increase, taxes. The 1975 legislature eventually agreed to put the property tax proposal on the March 9, 1976 presidential preference primary ballot. All of the districts except Northwest Florida would be authorized to levy up to 1.0 mills of property tax ($1.00 per $1,000.00 of assessed value). The Northwest District would be limited to only 0.05 mills, due to political maneuvering in the legislature at the time. Proponents of the tax proposal formed a group named FLOW ("Florida Love of Water") to stir up support.[26]

The *Orlando Sentinel* editorialized against the referendum. Some agricultural interests, like Florida Citrus Mutual, also opposed the proposal. They said they favored the general idea of levying property taxes but opposed this particular referendum proposal. The Hillsborough County Commission unanimously voted to oppose the referendum. Even some proponents of natural resource management recommended a No vote. Charles Lee, on behalf of the Florida Audubon Society, said, "We believe that the nebulous wording of the constitutional amendment which will be presented to the voters March 9 will serve to weaken water management under Florida's landmark Water Resources Act, should it be approved by the electorate." Audubon favored ad valorem taxation for the water management districts but also favored postponing the referendum so the wording of the constitutional amendment could be modified. Similar opposition came from Johnny Jones, executive director of the Florida Wildlife Federation. Art Marshall also opposed the referendum, believing that it would allow the existing districts to continue building too many structures, that the referendum was poorly worded, and that the beneficiaries of water management activities should pay for it.[27]

Governor Reubin Askew recommended that voters pass the constitutional amendment:

> I urge your support and vote for the water amendment. Water is the most important substance on earth. Life and growth are impossible without it. Florida, its farms, homes, and industries must have enough fresh water if we are to continue to have the quality of life that has made our State the envy of and magnet for people all over the world. There are those with the same commitment to conservation of our natural resources who oppose or would put off a vote for the water management amendment. They seek a more "perfect" instrument to get more local tax funds for northwest Florida for water management or prefer complete control and funding from Tallahassee.
>
> Of course, there are others who just object to governmental efforts to manage growth and our natural resources. I sincerely believe that the constitutional amendment proposed is needed now and that it will give a local tax base and revenue to go with State general revenue and any available federal funds for a balanced state-local water management program.
>
> So again I urge your support and vote on March 9 for the constitutional amendment.[28]

When the referendum votes were tallied, it received a 55 percent majority, marking the only time in Florida history that statewide voters approved ad valorem taxation. Of the state's 67 counties, only 18 voted for it by a majority but they were some of the most populous. No county north of Orange County voted for it. Around the state, the vote on the referendum split as follows:

SFWMD (region): 66% for, 34% against
SWFWMD: 56% for, 44% against
SJRWMD: 47% for, 53% against
SRWMD: 36% for, 64% against
NWFWMD: 35% for, 65% against[29]

The majority vote by the electorate put in place a stable financial foundation for the state's critically important water management system. Today, the proposed amendment would not have made it into the constitution because in 2006 Floridians amended their constitution to require that new amendments be favored by a majority of at least 60 percent.

In the relatively short period from 1971 to 1975, Florida had experienced a major drought, rewrote the fundamental water law of the state, created a statewide system of water management districts, and had given the new institutions (except the Northwest Florida Water Management District) a stable source of funding. Nelson Blake, in his classic 1980 book on the history of water management in Florida, called this period, "Florida Takes the Lead."[30] The enactment of the Florida Water Resources Act of 1972 marked a dramatic change in the state's approach to water management.

Would the new system meet the challenges of water management in Florida? Is the framework established in 1972 still capable of meeting the needs of the twenty-first century? Those are the questions addressed in the next chapters.

NOTES

1. Blake, Nelson. (1980) *Land into Water – Water into Land*, University Press of Florida, Gainesville, FL, p32

2. Derr, Mark. (1989) *Some Kind of Paradise: A Chronicle of Man and the Land in Florida*, William Morrow and Company, New York, p88

3. Gifford, John C. (1911) *The Everglades and Other Essays Relating to Southern Florida*, Everglade Land Sales Co., Kansas City, MO, p102, retrieved from http://digitool.fcla.edu/R/GA4MVUSK3EIJ8XK5PRLMELPGKB9Y45EITMTM4BU D8ESVHVJ8F8-02340?func = dbin-jump-full&object_id = 146571&local_base = GEN01-FIU01&pds_handle = GUEST, accessed 14 January 2011

4. [Florida] State Committee on Water Resources. (1945) *Report and Legislative Recommendations*, April 1, 1945, p3

5. Lodge, Thomas E. (1994) *Everglades Handbook: Understanding the System*, St. Lucie Press, Delray Beach, FL, pp172–173

6. Chapter 57-380, Laws of Florida

7. Oral History Program. (2006) Dale Twachtmann, 20 February

8. Oral History Program. (2000) Nathaniel Reed interview, 2 November

9. Burnett, Gene M. (1986) *Florida's Past: People and Events that Shaped the State*, Pineapple Press, Sarasota, FL, p106

10. Weitz, Seth A. (2007) "Bourbon, pork chops, and red peppers: Political immorality in Florida, 1945–1968," Ph.D. dissertation, Florida State University, Gainesville, FL, p208

11. Weitz, Seth A. (2007) "Bourbon, pork chops, and red peppers: Political immorality in Florida, 1945–1968," Ph.D. dissertation, Florida State University, Gainesville, FL

12. Weitz, op. cit. note 10, p208

13. Oral History Program, op. cit. note 8

14. Carter, Luther J. (1974) *The Florida Experience: Land and Water Policy in a Growth State*, published for Resources for the Future by The Johns Hopkins University Press, Baltimore, MD, p125

15. Governor's Conference on Water Management in South Florida. (1971) "Statement to Governor Reubin O'D. Askew," *Water Management Bulletin*, vol. 5, no. 3

16. Oral History Program. (2001) John DeGrove, 1 December

17. Smith, Zachary A., and Carol Johnson. (2008) "AMA reform: A political analysis," Appendix F in Sharon Megdal, Zachary A. Smith, Aaron M. Lien, and Carol Johnson (2008),

Evolution and Evaluation of the Active Management Area Management Plans, Arizona Water Institute, University of Arizona, Tucson, AZ, pp1–97

18. Oral History Program, op cit. note 16

19. Maloney, Frank E., Richard C. Ausness, and J. Scott Morris. (1972) *A Model Water Code, with Commentary*, University Press of Florida, Gainesville, FL

20. Oral History Program. (2005) Jacob D. Varn, 17 November

21. The designated members of the Task Force on Resource Management were John DeGrove (Florida Atlantic University), Art Marshall (University of Miami), Gilbert Finnell (Florida State University School of Law), Florida state representative Don Crane, Florida state senator Bob Graham, Bruce Johnson (Coastal Coordinating Council in the Florida Department of Natural Resources), Florida state representative Ray Knopke, Jack Maloy (Central and Southern Florida Flood Control District), Don Morgan (Central and Southern Florida Flood Control District), Richard Rubino (Florida State University Department of Urban and Regional Planning), Nils Schweizer (architect, Winter Park), Florida state senator Jack Shreve, Earl Starnes (Florida Department of Transportation), Homer Still (Bureau of Planning in the Florida Department of Administration), and Norm Thompson (Tampa Bay Regional Planning Council). This list is from Richard G. Rubino and Earl M. Starnes (2008), *Lessons Learned: The History of Planning in Florida*, Sentry Press, Tallahassee, FL, p227

22. Stephenson, R. Bruce. (1997) *Visions of Eden: Environmentalism, Urban Planning, and City Building in St. Petersburg, Florida, 1900–1995*, Ohio State University Press, Columbus, OH, p147

23. Oral History Program. (2004) Derrill McAteer, 22 June

24. Lewis, Phil [Florida state senator]. (1975) "Water and power," *St. Petersburg Times*, 9 June

25. Oral History Program. (2007) Don Feaster, 24 January

26. "Flow, Inc., seeks water district tax." (1975) *Ft. Lauderdale News*. 13 December

27. Citrus Mutual: Fisher, Jim. (1976) "Citrus Mutual urges defeat in water district tax vote," *Tampa Tribune*, 5 March; Hillsborough County: Friedman, Barry. (1976) "Spicola urges weighted water vote law," *St. Petersburg Times*, 4 March; Charles Lee (1976): "Letter to editor," *St. Petersburg Times*, 7 March; Art Marshall (1976): E. D. Vergara, Southwest Florida Water Management District memorandum, "Major points against the referendum by Dr. Art Marshall and Frank Caldwell," January 26, 1976

28. Askew, Reubin [former Florida governor]. (1976) "Water management taxes necessary," *Florida Conservation News*, February 1976, p4

29. "Referendum votes by district," Florida Water Law Collection Digital Archive, http://ufdcweb1.uflib.ufl.edu/ufdc/?a = flagua&m = hdFC&i = 51599, accessed 13 January 2011

30. Blake, op. cit. note 1, p223

CHAPTER THREE

The Responsibilities of Water Management

A ll states have water management agencies of one kind or another. Most of these agencies in other states focus on single issues like water supply. Florida is remarkable for being the only state with a statewide system of regional water agencies with comprehensive water management responsibilities.

THE NEED FOR COMPREHENSIVENESS IN WATER MANAGEMENT

Water management district responsibilities are very broad — much broader than the water supply or flood control mission for which they are widely known. Lake Okeechobee is illustrative. The big lake is managed, but only in part, to provide water supply for southeast Florida. A narrowly focused regional water supply agency could take on that task as a large mission, all by itself. However, Lake Okeechobee also has to be managed to prevent the type of catastrophic flooding that occurred there twice in the early twentieth century. Nor do the demands for effective water management end there. The lake also has significant water quality problems from excessive nutrients.[1] Okeechobee also is a famed recreational resource and a significant habitat for threatened and endangered species. The problems and opportunities are multiple and interconnected.

The risk of comprehensiveness is that the combining of responsibilities in an agency might prevent full and open airing of the genuine conflicts among multiple responsibilities because discussions are confined within a single organization. That danger is most acute when the district is responsible both for issuing permits for water withdrawals and for promoting water supply development.

Chapter 8 on the St. Johns River Water Management District provides an example of how that conflict of roles can endanger the public's trust in a water management agency.

Florida's system of regional water management districts is the lineal descendant of more than a century of attempts to exploit and manage water resources. Since the 1972 water management reforms described in the previous chapter, the system has evolved substantially while retaining the overall structure that was put in place that year. When enacted, the state's framework for water management occupied 19 pages of the statute book; in 2010 it took more than 150 pages. In the fundamental Declaration of Policy of the modern Water Resources Act, water management districts are assigned a long array of important tasks:

(1) The waters in the state are among its basic resources. Such waters have not heretofore been conserved or fully controlled so as to realize their full beneficial use.

(2) The department and the governing board shall take into account cumulative impacts on water resources and manage those resources in a manner to ensure their sustainability.

(3) It is further declared to be the policy of the Legislature:

    (a) To provide for the management of water and related land resources;

    (b) To promote the conservation, replenishment, recapture, enhancement, development, and proper utilization of surface and ground water;

    (c) To develop and regulate dams, impoundments, reservoirs, and other works and to provide water storage for beneficial purposes;

    (d) To promote the availability of sufficient water for all existing and future reasonable-beneficial uses and natural systems;

    (e) To prevent damage from floods, soil erosion, and excessive drainage;

    (f) To minimize degradation of water resources caused by the discharge of stormwater;

    (g) To preserve natural resources, fish, and wildlife;

    (h) To promote the public policy set forth in s. 403.021;

    (i) To promote recreational development, protect public lands, and assist in maintaining the navigability of rivers and harbors; and

    (j) Otherwise to promote the health, safety, and general welfare of the people of this state. (Section 373.016, Florida Statutes).

And that is not all. Water touches everything in Florida, and the water management system is correspondingly complex. Other sections of contemporary Chapter 373, Florida Statutes, give many other tasks to the districts. For example, the long list of specific responsibilities assigned to the water management districts includes the following summary list:

- Identify enough water supplies to meet all reasonable beneficial needs for the next twenty years (Section 373.0361(2)(a)).
- Help to assure that "Sufficient water [is] available for all existing and future reasonable-beneficial uses and the natural systems, and that the adverse effects of competition be avoided (373.0831(2)).
- Develop a funding strategy to pay the cost of all of the recommended projects (373.0361(2)(b)).
- Assist local governments in the development and revision of their comprehensive plans (373.0361(8)).
- Establish minimum flows and levels for surface and groundwater resources (373.042 and 373.0421).
- Administer water use, well construction, and environmental resource permitting systems (Parts II, III, and IV of Chapter 373).
- Adopt rules for any action that affects interested parties (373.113), consistent with the many procedural and substantive requirements of the Florida Administrative Procedures Act (Chapter 120).
- Acquire and manage land and facilities (373.139).
- Act as local sponsor for projects with the Army Corps of Engineers, including the world's largest environmental restoration project in the Everglades (373.1501, 373.4592, 45926, 470, 472, 503).
- Promote Florida-Friendly landscaping (373.185).
- Plan and help fund the development of "alternative water supplies" (373.196).
- Assist in the development of regional water supply authorities (373.1961, 1962, and 1963).
- Implement a large land acquisition program (373.199).
- Plug abandoned artesian wells (373.207).
- Implement a water use permitting program, with many conditions and exceptions (373.216 and other provisions of Part II of Chapter 373).
- Assist in the statewide water conservation program for public water supply (373.227).
- Promote the reuse of reclaimed water (domestic wastewater) (373.250).
- License water well contracting businesses (373.323).
- Manage a comprehensive stormwater and wetland permitting program (Part IV of Ch. 373).
- Coordinate district permitting with the related wetland programs of the federal government (373.4143).
- Inspect stormwater management systems as they are constructed (373.423).
- Facilitate the use of "mitigation banks" to offset the impacts of wetland permitting (373.4135 and 4136 and 4137).
- Direct the emergency operation of any stormwater management system or reservoir so as to protect life and property (373.439).

- Conduct annual budget, including public workshops and hearings (373.536, 539).
- Sell and manage bonds (373.563 and others).
- Assist the state in issues of interstate water management.

There are additional unique responsibilities assigned to the South Florida Water Management District, including the following:

- Carry out the Comprehensive Everglades Restoration Plan (Section 373.26(8)).
- Implement restoration activities in the "Northern Everglades and Estuaries" region (Lake Okeechobee, Caloosahatchee, and St. Lucie Rivers) (373.4985).

Each of these responsibilities is a management program by itself. They require large budgets and adequate coordination in the work performed every day by the water management districts. That coordination is possible because the long list of tasks assigned to the water management district is guided by a smaller set of foundational principles, described next.

WATER IS A PUBLIC RESOURCE, NOT PRIVATE PROPERTY

Until 1972, state law referred to a supposed "property right" to water (Section 373.101, F.S.). The Model Water Code argued instead that water should be held as a public trust:

Water, as a resource, cannot be described as being permanently situated within any particular boundaries. Since no citizen can permanently own the state's water resources or totally deny other citizens the right to use them, water resources do not fall within the classic definition of property rights. Each citizen's right in the water can best be described as a right to common use of a resource to be used by all but owned by no one.[2]

The Water Resources Act, adopted by the Florida legislature, follows that general principle in the Code but did not incorporate verbatim the specific language about "property of the state":

Because water constitutes a public resource benefiting the entire state, it is the policy of the Legislature that the waters in the state be managed on a state and regional basis. Consistent with this directive, the Legislature recognizes the need to allocate water throughout the state so as to meet all reasonable-beneficial uses. (Section 373.016(4)(a))

The Supreme Court of Florida, interpreting the Water Resources Act, stated definitively that "the ownership of land does not carry with it any ownership of vested rights to the underlying ground water not actually diverted and applied to beneficial use." Beyond that, the court found that the "only water rights that one has are those rights to withdraw water subject to the terms and conditions of a permit issued by a water management district."[3]

A more vivid explanation of water rights was offered by Jacob Varn, a water lawyer and former secretary of the Florida Department of Environmental Regulation:

> You know, I always try to make a distinction, but you often hear the debate about who owns the water? To me that's akin to who owns the air, or who owns the deer or the birds. The only thing I know is, once you capture it, I know who owns it. Before that, it's the subject of a lot of debate.[4]

"REGULATED RIPARIANISM"

The classic division in types of water law in the United States is between the Western system of "prior appropriations" and the Eastern system of "riparian rights." In its origin and in its purest form, the Western system is oriented toward private property in water itself. The first party to withdraw water from a water source has priority over later users; subsequent claims to water follow according to when they began withdrawing water. In contrast, eastern states historically tended to associate water withdrawal rights with the ownership of land; water use also had to be "reasonable." In recent decades, both systems have tended to incorporate features not present originally, such as water use permitting, watershed management, factors of the public interest, and protection of minimum flows and levels.[5]

Florida water law is clearly on the riparian side of the water law river. It has a form of "regulated riparianism," rather than relying upon outdated connections between water use and property ownership. The American Society of Civil Engineers counted 17 eastern states that also have enacted a form of regulated riparianism for either ground or surface waters, for at least part of the state.[6] Although they follow the regulated riparianism doctrine, none of these states has anything like Florida's complex system of integrating the basic law of water use permitting with other water management responsibilities.

PERMITS TO USE THE WATER

In Florida, all water use, except for individual domestic use by households, requires a permit from a water management district. The permits are not perpetual. Except for public water suppliers, they usually cannot exceed 20 years. Most importantly, there is no legal "right" to use water independently of a water use permit, as an administrative law judge ruled in 2008 in regard to a permit application for bottling water:

> Water rights are not acquired by plans, hopes, desires, or interests, but by legal diversions of water to reasonable beneficial uses. Petitioners' reference to a "right to use groundwater held in trust for them and their residents" has no support under Florida law.[7]

The water use permitting system in Florida is as comprehensive as that found in any state. Nineteen of the 28 states east of the Mississippi River require some form of water use permit. In most of those states, however, permits are required for ground water, or only for surface water, or only for certain regions.[8] Even though many states have, in principle, a system of regulated riparianism, it is often not very extensive in practice. Only a few states have comprehensive water use permit systems.[9]

WATER WITHDRAWALS MUST MEET THE "THREE-PART TEST"

The specific requirements to obtain a water use permit are at the foundation of Florida's water management system. Applicants for water withdrawal permits in Florida must satisfy three separate criteria:

> To obtain a permit pursuant to the provisions of this chapter, the applicant must establish that the proposed use of water:
>
> (1) Is a reasonable-beneficial use as defined in s. 373.019;
> (2) Will not interfere with any presently existing legal use of water; and
> (3) Is consistent with the public interest. (Section 373.223, F.S.)

The statute defines what is meant by "reasonable-beneficial use":

> "Reasonable-beneficial use" means the use of water in such quantity as is necessary for economic and efficient utilization for a purpose and in a manner which is both reasonable and consistent with the public interest. (Section 373.019(16), F.S.)

The requirement for "efficient" utilization of water (or water conservation, as it is also called) is at the very center of the reasonable-beneficial use test in permitting. Interestingly, the public interest test occurs both as one of the three criteria for permit issuance and then again as part of the definition of "reasonable-beneficial use." Neither the Model Water Code nor the Water Resources Act provides guidance on the meaning of "interfere with any presently existing legal use of water" or "consistent with the public interest." In a water bottling case, Administrative Law Judge Bram Canter found in 2009 little difference between the two usages of the "public interest" term, concluding that:

> It appears to do no more than give consideration of the public interest a prominent place in water use permitting, on the same footing as reasonable-beneficial and avoiding interference with existing water users. The third prong does not expand the public interest inquiry beyond water resource-related issues.[10]

If that view prevails, water management districts cannot consider the broad public interest in making water use permit decisions but only how water resources are affected.

Three law professors at the University of Florida, where the Model Water Code was written four decades ago, distributed a defense of the Water Resource Act in 2008, together with their set of proposed reforms. Professors Angelo, Hamann, and Klein recommended that water resource sustainability should be the "benchmark of the public interest" and that several additional specific factors could be included in a list of considerations to interpret the public interest.[11]

"EXCLUSIVE AUTHORITY"

Not only are the water management districts directed to have a water use permitting program, they are the only government in Florida that can do so. The legislature provided that the Water Resources Act "preempts" any other regulation of water use (Section 373.217, F.S.). Local governments therefore are proscribed from regulating water use. They cannot put conditions on withdrawals of water for municipal, agricultural, or any other use. For example, this precludes local governments from prohibiting the transport of water outside their jurisdictions or forbidding unpopular uses such as water bottling. (Local governments sometimes try to work around this statutory limitation on their authority by focusing on the land use implications of the activity that withdraws water.)

ONE LAW FOR SURFACE AND GROUNDWATER

The basic principles of Florida water law today make no distinction between surface and groundwater. Many other states have different legal doctrines for water above and below ground. In Arizona, for example, Bonnie Colby noted that the "scientific recognition of the connectivity between groundwater and surface water has not been integrated in Arizona water law, which remains bifurcated."[12] It makes little legal or hydrological sense to have separate doctrines for surface water and groundwater. This is especially true in Florida where surface water and groundwater features are so closely connected and interchange so freely. When a stream receives groundwater flow seeping into it from adjoining lands, it is the same water in the river and out of the river. Florida is fortunate that a single legal doctrine applies to both surface water and groundwater, unlike many other states.

CENTRALIZED (WITHIN REGIONS)

Observers of federal water policy have long lamented that national water policy is highly fragmented. For example, Jonathan Deason and colleagues noted water policy involvement by "at least thirteen congressional committees, twenty-three congressional subcommittees, eight Cabinet level departments, six independent agencies, and two White House offices. To further complicate water resource policy issues, those federal entities with authority over water resource planning are not the same entities that have jurisdiction over the funding for water-related projects."[13] Unlike the federal system, and unlike many states, there is substantial consolidation of responsibilities within each of Florida's five water management districts.

Although other agencies in Florida, such as "water control districts" (Chapter 298, F.S.), retain some local water management responsibilities, the water management districts overwhelmingly have a lead role. Few states have adopted substate agencies for regional or watershed level water management. One study says that "Two states – Nebraska and Florida – are pioneers in managing water across the state at a scale larger than the county level but smaller than the state."[14] Water managers in Florida do not have to face the same degree of splintered management that is the norm in many other states.

POWERFUL APPOINTED BOARDS

The governing boards of the water management districts are not elected, but are appointed by the governor and confirmed by the state senate. Governing board members serve four-year terms, work hard, and often devote many hours (or even days) a month to their duties. For the larger districts especially, each month brings not only regular governing board meetings, but also other workshops and events in the district. Board members receive no pay except travel expenses. Each board has a mix of members who must reside in different watersheds or counties in the district, as well as a few members who serve at large. The board sets district policies, which are carried out by an executive director whom they choose. The executive director is a full-time paid employee of the district.

At times, because the governing boards are appointed rather than elected, their ability to levy ad valorem property taxes has been called "taxation without representation." Appointment, rather than election, was recommended in the Model Water Code. The appointment method was also recommended decades later by the American Society of Civil Engineers:

> The members of an Area Water Board could be elected by the general electorate within the Area or perhaps in some manner designed to represent water users within the Area, or appointed by local units of government within the Special Water Management Area, or even appointed by various departments of the State Government. The Regulated Riparian Model Water Code rejects these alternatives because it makes it too likely that the Area Water Board would see its function as serving the short-term needs of those who elect or appoint it rather than as implementing the State's regulatory policies relative to the waters of the State.[15]

Perhaps the most fundamental response to a concern about the taxation role of the water management districts is that 55 percent of the voters of Florida put that system into their state constitution. It is also true that the governor, who appoints board members, is elected. The legislature, which has capped the millage level of the water management districts at a level lower than the constitutional maximum, is also elected. All of the governing board members are confirmed by the Florida senate. Executive directors must be confirmed by the senate after a change of governors. The legislature specifies the duties of the water management districts, receives multiple reports from them each year, and decides on the level of funding from the state for water management purposes, which is in addition to property taxes levied by the districts.

Appointed governing board members would appear more likely to be interested in the needs of the entire region than officials elected from small fractions of it. If governing board members were elected by designated subdistricts, they would tend to represent the interests of their respective subdistricts. If they were elected districtwide, surely the populated urban regions would end up selecting the great majority of governing board members. Rural and agricultural interests, now heavily represented on governing boards, would seem likely to become a permanent minority. The Florida Farm Bureau recognizes that the current appointment method favors their industry and opposed in early 2011 any change from the appointment process.[16]

The districts clearly are not exempt from the political process. Since 1997, the legislature has also injected both the governor and itself into the selection by the governing board of an executive director:

> The appointment of an executive director by the governing board is subject to approval by the Governor and must be initially confirmed by the Florida Senate. The governing board may delegate all or part of its authority under this paragraph to the executive director. The executive director must be confirmed by the Senate upon employment and must be confirmed or reconfirmed by the Senate during the second regular session of the Legislature following a gubernatorial election. (Section 373.073(1)(b), F.S.)

This senatorial review of executive directors was placed into statute by the Republican-majority legislature in advance of the likely election of Republican Jeb Bush to the governor's office. Such senate "reconfirmation" undercuts the authority of a governing board to perform its own supervision of a water management district. Other political maneuvers also occur. In 2008, for reasons that senators would not state, the executive director of the Southwest Florida Water Management District who had served satisfactorily for five years, and was endorsed by his governing board, did not receive senate confirmation and was held over until the next year for final approval.[17]

The only qualifications for appointment to a governing board are the ability to secure the endorsement of a governor and the following:

> Membership on governing boards shall be selected from candidates who have significant experience in one or more of the following areas, including, but not limited to: agriculture, the development industry, local government, government-owned or privately owned water utilities, law, civil engineering, environmental science, hydrology, accounting, or financial businesses. (Section 373.073(2), F.S.)

The disciplines most often absent from a governing board are environmental science and hydrology. Only people with influence in a community are able to secure an appointment by an elected governor. They tend to be safe, conservative choices for these important positions. Most of them fit within George Gonzalez's definition of an "economic elite" who "benefit from economic growth in a specific locality or region."[18] Some of the appointed governing board members have business interests in matters regulated by the water management district. Board members recently included an operator of a dairy farm in north Florida, an employee of a large water utility in Jacksonville, a vice-president of the largest sugar grower in the Everglades Agricultural Area, the operator of a citrus grove, and a vice-president of Disneyworld Resorts. In cases of conflict of interest that would cause "special private gain," board members should abstain from voting and then file a form stating the reason for the abstention.[19] However, they can still participate in the discussions leading up to a vote and they can vote on any matter not leading to their "special private gain." For example, they can vote on matters affecting the type of business in which they are employed but should not vote on a water use permit for their direct employer.

The appointments tend to lag behind changing patterns of American diversity. For example, no woman was appointed to the SFWMD governing board until Jeanne Bellamy's term began in 1979. Pamela Thomas-Brooks was the first African-American appointed to SFWMD in 2001. The Southwest Florida Water Management District's first woman chair was selected in 2005; St. Johns River Water Management District did not name one until 2008. In 2010, 80 percent of the five district's governing board members were men. There is usually only a handful of African-Americans.

Estus Whitfield, who played a role in advising three governors on water management, felt "absolutely" that appointed board members were better than elected ones, that he would "defend that to the end":

> I think all the governors that I've been familiar with and worked for have been extremely conscientious in appointing board members to water management boards because the governors have always known, going from Askew to Graham to Martinez to Chiles to Bush, that the water management districts are probably the most significant entities in the state of Florida as far as affecting the physical and growth management aspects of the state.[20]

LARGE BUDGETS

A survey of eastern states' water management practices noted that the authorizing statutes sometimes seemed better in theory than in practice. Implementation funding often fell short of needs:

> Adequate funding will be a key to any system which moves toward more regulatory control of water resource use and development. Lack of adequate

funding will not only convey a false impression that water resources are actually being protected and effectively utilized, it will also magnify the potential for "bad" bureaucratic decisions which can affect a state's economy and water resources as negatively as the deficiencies of reliance on the common law.[21]

No district has ever levied the full mill that the constitution allows. In the South Florida Water Management District, for example, the assessed millage in 2010–2011 is 0.624 mills (62.4 cents per $1,000 of assessed value.) A home with a taxable value of $150,000 (assessed by the county property appraiser at $200,000, less the $50,000 homestead exemption) would pay $93.60 a year to the district. With assistance from state appropriations, the overall district budgets are large. In recent years, the combined budgets of the five districts ranged from $1.2 billion to $2.3 billion a year.

LARGE STAFFS

At the state level, the Department of Environmental Protection has about 3,600 employees. Less than 1,000 are involved in any kind of water management activity, ranging from regulating drinking water and wastewater systems to restoring the state's beaches. Total employees of the water management districts are about 3,400 – many more than employed by state government for water management purposes. Water management district staff, on average, are also better compensated than state employees.

The level of staffing of the water management districts is an anomaly in a state with a strong belief in being cheap. Florida was tied for the lowest number of full-time state employees per capita (118 employees per 10,000 people, compared to the national average of 216). Florida also has the lowest state payroll per resident of all 50 states ($38 per state resident of Florida; national average is $72.)[22] Having the lowest number of state workers does not prevent calls for making the number even smaller. Governor Rick Scott, elected in 2010, called for reducing the number of state employees by 7 percent in his first budget proposed in early 2011.[23] The existence of the water management districts, and their ability to levy property taxes, has spared the legislature the expense and political trouble of raising state taxes for that purpose. Scott's deep-seated belief in the need to cut the size of government also led him to propose a two-year 25 percent cut in the property taxes levied by the water management districts.[24] Although Scott campaigned on a platform of reducing taxes and the scale of government, it was not until he released his state budget proposal a month after inauguration in January 2011 that this idea was promoted. Suddenly reducing water management districts revenues by this amount would disrupt ongoing responsibilities and long-range projects. In early 2011, it is impossible to tell if this is a serious proposal and indicative of the new Governor's long-range plan for natural resource management in his adopted state or just political positioning.[25]

WATER SHORTAGE MANAGEMENT

In times of water shortage, the districts are granted very broad powers to respond. In "normal" water shortage conditions, a district can "impose such restrictions on one or more users of the water resource as may be necessary to protect the water resources of the area from serious harm." (Section 373.175(2), F.S.) Under "emergency" conditions, statutory powers are even more sweeping, including "apportioning, rotating, limiting, or prohibiting the use of the water resources of the district." (Sections 373.175(4) and 373.246(7), F.S.) Unlike the Central and Southern Flood Control District in 1971, the water management districts today are equipped with wide powers to respond to water supply crises.

BASIN BOARDS

Only two water management districts have additional "basin boards" (also appointed by the governor) that report to the main governing board of the entire district. The South Florida Water Management District has a single additional board for the western portion of their district. However, the Southwest Florida Water Management District has seven separate basin boards, each with their own appointed members and an ability to levy property taxes. District staff spend a significant fraction of most months at SFWMD preparing for basin board meetings and following up on the results.

"LOCAL SOURCES FIRST" AND LONG-DISTANCE TRANSFERS

Florida water law discourages long-distance transport of water like the California Aqueduct, which moves water from the northern end of the state to the southern end. The "Local Sources First" policy requires water users to first try to obtain water in their own geographic area before proposing to get water from other counties, let alone from other water management districts. A proposal in 2003 to study the feasibility of a "statewide water distribution system" led to public hearings in north Florida, with up to 1,000 opponents of the idea vehemently objecting. Chapter 373 includes several policies on this subject, including:

> [...]the Legislature directs the department [of Environmental Protection] and the water management districts to encourage the use of water from sources nearest the area of use or application whenever practicable. Such sources shall include all naturally occurring water sources and all alternative water sources, including, but not limited to, desalination, conservation, reuse of nonpotable reclaimed water and stormwater, and aquifer storage and recovery. (Section 373.016(4), F.S.)

The policy has broad support in most of the state and was not opposed in the recommendations from the 2008 Water Congress meeting. The "Local Source

First" policy amounts to a determination to live within our means in self-sustaining watersheds.[26]

WATER PLANNING

A 2006 survey of state water plans sponsored by the American Society of Civil Engineers found "The principal weakness in many state water plans is the continued compartmentalization of aspects such as water quantity and quality, and groundwater and surface water."[27] There is in fact an indivisible hydrological and management relationship between water supply, water quality, natural systems, and floodplain management. The original Model Water Code recommendation for a "State Water Use Plan" was enacted in the Water Resources Act of 1972. Over time, in recognition of the interrelated responsibilities of water management, the state plan was broadened beyond only "water use" to full, multipurpose water management. Governor Bob Martinez created by executive order in 1989 the Governor's Water Resource Commission, which reviewed the water management system to see if improvements were desirable. They recommended that a "State Water Management Plan" and related "District Water Management Plans" be prepared by 1994. The statute was realigned to promote a comprehensive planning approach at both the state and district levels.

Still, aside from the Water Resource Implementation Rule part of the Florida Water Plan set out in Section 373.036(1)(d) above, the Florida Water Plan has had very little influence. John DeGrove (the "father" of Florida growth management) believed that the promise of the Florida Water Plan (and other state agency plans) was undercut by the failure of the state to deliver the billions of dollars necessary for effective growth management.[28] Another factor in planning failure is that much of the day-to-day work of the Florida Department of Environmental Protection is focused on the tens of thousands of permits issued by the agency each year, rather than on long-range planning.[29] Andrew Dzurik, the author of a well-regarded text on water resource planning, is right in observing that "the day-to-day pressure of regulatory decisions is simply a foreign environment for long-range planning."[30]

THE WATER RESOURCES IMPLEMENTATION RULE

The Water Resources Act requires the Department of Environmental Protection to oversee the five water management districts. In the early 1980s, the department adopted what was then called the "Water Policy" rule (Chapter 62-40, Florida Administrative Code) to provide some structure for that guidance. It serves as the department's tool for amplifying the guidance in the underlying statute. Periodic updates to the rule help address contemporary water management needs. In 2007, for example, the department added a new section to the rule setting out general policy on the implementation of "reservations" of water. This was necessary, as discussed in the Reservations section of this chapter, to deal with the need to adopt a series of reservations of water associated with the Comprehensive Everglades

Restoration Plan. Amendments to the rule are usually opposed by development interests and occur only every few years.

REGIONAL WATER SUPPLY PLANNING

One type of regional water supply planning is conducted by counties and cities that voluntarily form a regional water supply authority. However, only a few such authorities have been formed in Florida since 1974, despite a consensus that they have been relatively successful in meeting the water supply needs of their member governments.[31] Another type of directive on regional water supply planning was added by the legislature in 1997.[32] The water management districts must produce district-wide assessments to determine which parts of each district do not have adequate supplies of water planned for the next 20 years. For the areas that did not have 20-year supplies determined, the districts were required to prepare complete regional water supply plans. The plans now identify sources of supply that exceed growing demands over every two-decade period. Local governments are encouraged, but not required, to follow the regional water supply plan. The whole planning process was set on a 5-year update cycle, with the most recent updates to be completed across the state in 2010 and 2011.

MINIMUM FLOWS AND LEVELS

Water use permits authorize the removal of water from natural systems. At some point, if too much water is withdrawn, the water body is damaged. Taking too much water via water use permitting can lead to water resource destruction. A river that is denied the water it needs will not have adequate habitat for aquatic organisms. Estuaries, which depend on maintenance of freshwater flows, will convert to saltwater habitats. Coastal wellfields, if water levels are driven too low, will invite in salty ocean water and have to be closed (or groundwater supplies will be forced to undergo expensive water treatment).

To prevent these effects, the Water Resources Act requires the water management districts to establish "minimum flows and levels" (MFLs) for surface water and groundwater resources. The minimum flow is the limit at which further withdrawals from a surface water body would be "significantly harmful to the water resources or ecology of the area," while a minimum level for groundwater attempts to prevent harm to the "water resources of the area." The natural hydrologic regime of water bodies determines ecological health. The desired minimum flows are not just the extreme low flows allowable, because water body health depends on the maintenance of the right hydrologic regime at all flow levels. Since 1997, the water management districts have established more than 300 separate MFLs. Additional MFLs are established each year in accordance with priority lists submitted by the district to DEP for approval.

The establishment of minimum flows and levels is one of the most powerful tools in the Water Resources Act to maintain water resource sustainability. A law enacted by the 2010 Legislature requires that all agency rules be ratified by them before they can go into effect. The Legislature has no special expertise in determining waterbody levels of harm and the law poses a very serious obstacle to future adoption of minimum flows and levels.

RESERVATIONS

Minimum flows and levels are one process to protect natural systems against overwithdrawals, but are not the only tool available to prevent harm to water bodies. The Model Water Code and the Water Resources Act include other provisions for this purpose, including "reservations of water." The Water Resources Act allows, but does not require, that reservations be set:

> The governing board or the department, by regulation, may reserve from use by permit applicants, water in such locations and quantities, and for such seasons of the year, as in its judgment may be required for the protection of fish and wildlife or the public health and safety. Such reservations shall be subject to periodic review and revision in the light of changed conditions. However, all presently existing legal uses of water shall be protected so long as such use is not contrary to the public interest. (Section 373.223(4), F.S.)

The effect of reserving water from water use permitting is similar to that of establishing MFLs. However, MFLs are to prevent "significant harm to water resources or ecology of the area," whereas reservations are for the "protection of fish and wildlife or the public health and safety." Reservations matter more now than at any time since the creation of the Water Resources Act of 1972 because the federal law authorizing the Comprehensive Everglades Restoration Plan allows individual project components to be built only if water for the natural system is reserved or otherwise protected from allocation.[33] Just as the 2010 law about legislative ratification of rulemaking will make it harder to adopt minimum flows and levels, it will make it much more difficult and politicized for a water management district to adopt a reservation.

The South Florida Water Management District, as the district partner with the federal government on the Comprehensive Everglades Restoration Plan, is now establishing a series of water reservations.

LOOSE SUPERVISION

It is always a question of who is "in charge" of the water management districts. The Water Resources Act of 1972 gives the Department of Environmental Protection (formerly the Department of Environmental Regulation) "general supervisory authority" over the water management districts, as well as several specific supervisory instruments. The extent of general supervision is not defined and sometimes it is more theoretical than real.[34] One former district executive director remembers that, in the 1970s, some of the districts went around the back of the first secretary of the Department of Environmental Regulation and tried to "cut his throat."[35] A governing board chair remembers also thinking of the department staff person at water management districts meetings as a "spy."[36] Those days of direct antagonism have receded but are not entirely gone.

Nonetheless, the supervisory responsibility of DEP is very challenging. After all, both the secretary of the Department of Environmental Protection and every member of the governing board that is to be supervised are appointed by the governor. The secretary does not appoint the members of the boards and does not have the power to dismiss either them or the executive directors. Moreover, the governing boards have independent powers to assess property taxes and secure the revenues they deem necessary. The DEP is dependent on annual appropriations from the legislature and does not set the tax levels the governing boards assess.

Nonetheless, DEP is still required to exercise general supervisory authority over the water management districts (Section 373.026(7), F.S.) The general supervision by DEP is supplemented by a variety of specific tasks. For example, the department:

- Reviews and approves WMD priority lists for establishing minimum flows and levels (MFLs). DEP also is actively involved with some individual MFLs.
- Reviews WMD rules for consistency with the DEP Water Resource Implementation rule. (Chapter 62-40, F.A.C.).
- Reviews and comments on districts' water resource development work programs.
- Develops, in cooperation with the districts, the detailed format and guidelines for regional water supply plans.
- Prepares and distributes annual reports on the status of regional water supply planning throughout the state.
- Coordinates, with the districts, the state's response to droughts.
- Coordinates the Conserve Florida Water program, including the statewide water conservation clearinghouse, with funding assistance from the districts.
- Coordinates statewide policy development on rules for water use permitting with the water management districts.
- Coordinates the implementation of the Environmental Resource Permitting program, with the department taking the lead on some kinds of permitted activities, while the districts are primarily responsible for others.

The nature of the supervisory relationship will always have some contentious events because the state and regional water management agencies have overlapping responsibilities. Governors always have great influence over the water management districts. In addition to specific statutory powers, governors are the state's political leader and they appoint district governing board members and the Secretary of the Department of Environmental Protection. However, Governors must be careful not to supervise the districts so closely that they afoul of a tax prohibition in the state constitution. Section 18 of Article VII of the constitution forbids a state property tax:

> No tax shall be levied except in pursuance of law. No state ad valorem taxes shall be levied upon real estate or tangible personal property.

Despite this constitutional problem, the 2011 Legislature adopted for the first time direct limits on WMD budgets and taxes. If the Governor supervises the water management districts very closely, specifying their tasks and priorities, etc., that would amount to determining the ad valorem tax level necessary to perform duties. This incursion into district responsibilities would have an added accountability disadvantage. If the Governor is deeply involved in the supervision of the water management districts, final accountability would be diffused because it is divided between the Executive Director, the Governing Board, the Secretary of the Department of Environmental Protection, and the Legislature. The 2011 Legislature also muddied the responsibility for WMD supervision by imposing various budget restrictions and requiring the districts to have certain actions approved by the Legislative Budget Commission. If everyone is in charge, no one is in charge.

At times, proposals arise for a statewide body of some kind to oversee the water management districts rather than the Department of Environmental Protection. Staff of a Florida senate committee produced such a recommendation in 2009. They called for a "central regulatory commission that oversees Florida's water resources and supply development."[37] Floridians generally prefer their decentralized system of water management districts and fear what would happen if a vigorous state body could direct water to be piped from one region to another.

BUILDING SUSTAINABILITY?

The sprawling legal, financial, regulatory, and planning features of the Florida water management system have grown by accretion since 1972. Many functions and responsibilities have been added to the original Water Resources Act of 1972 without any comprehensive reconsideration of the basic structure. The result is that the current system is not clearly oriented toward long-term sustainability. This lack of clear focus on sustainability, on doing the "right" thing, is true even though the statute directs the water management system to "take into account cumulative impacts on water resources and manage those resources in a manner to ensure their sustainability." (Section 373.016(2), F.S.) That general statement about sustainability is undercut by the additional legislative directive that water management achieve some impossible things:

> Sufficient water [should] be available for *all* existing and future reasonable-beneficial uses and the natural systems, and that the *adverse effects* of competition for water supplies be avoided. [emphasis added]

It is not possible to make sufficient water available for *all* water uses, when many of those uses contribute little to Florida's economy or they are very wasteful. Only "reasonable-beneficial" uses should be allowed, which would mean not allocating water to trivial, harmful, or inefficient uses. A better view of the responsibilities of water management districts was given in 1978 by Derrill McAteer, the chair of the governing board of the Southwest Florida Water Management District:

> We are charged in the water management districts with the responsibility of protecting the resource; we do not necessarily represent the people. I don't represent people. I represent the resource and my position is appointed. My legislatively-named responsibility is to protect the resource, which I will do to the best of my ability.[38]

NOTES

1. There was a period of some confusion after creation of the Water Resources Act of 1972 about whether the new water management districts could regulate water quality. In 1975, the state's attorney general opined that the districts should not permit an activity that "unlawfully lowers water quality" but that it might be wise for the districts to request specific authority from the Florida Department of Pollution Control (predecessor to the Department of Environmental Protection) to act as their agent in water quality regulation. Attorney General Opinion AGP 75-16, "Flood control district and water quality," January 29, 1975.

2. Model Water Code, pp83–84

3. Committee on Natural Resources, Florida House of Representatives. (2003) *Water Rights in Florida: Interim Report*, December 2003, p6

4. Oral History Program. (2005) Jacob D. Varn, 17 November

5. Sherk, George William. (2003) "East meets west," *Water Resources Impact* (American Water Resources Association), vol. 5, no. 2, p8

6. American Society of Civil Engineers (ASCE) (2004) *Regulated Riparian Model Water Code*, Section 2R-2-24, ASCE/EWRI 40-003, ASCE, Reston, VA, p ix

7. Florida Division of Administrative Hearings. (2008) *City of Groveland and Lake County v. Niagara Bottling Company and St. Johns River Water Management District*, Judge Bram D. E. Canter, Case No. 08-4201, Order of Dismissal, September 11, 2008

8. Rusert, Whitney, and Ronald Cummings. (2004) *Characteristics of Water-Use Control Policies: A Survey of 28 Eastern States*, Water Policy Working Paper No. 2004-001, North Georgia Water Planning and Policy Center, p3

9. Tarlock, A. Dan. (2004) "Water law reform in West Virginia: The broader context," *106 West Virginia Law Review*, 495-538, p535

10. Florida Division of Administrative Hearings. (2009) *City of Groveland v. Niagara Bottling Company, LLC, and St. Johns River Water Management District*, Judge Bram D. E. Canter, Case No. 08-4201, Recommended Order, August 7, 2009

11. Angelo, Mary Jane, Richard C. Hamann, and Christine A. Klein. (2008) *Reforming the Florida Water Resources Act of 1972: A Sourcebook*, University of Florida Law School, Gainesville, FL, pp1–53. The authors also published a modified version as "Modernizing water law: The example of Florida" in *Florida Law Review*, vol. 61, no. 3 (2009), pp1–73

12. Colby, Bonnie G., Katharine L. Jacobs, and Dana R. Smith. (2007) "Lessons for semiarid regions facing growth and competition for water," in Bonnie G. Colby and Katharine L. Jacobs, (2007) *Arizona Water Policy: Management Innovations in an Urbanizing, Arid Region*, Resources for the Future, Washington, DC, p220

13. Deason, Jonathan P., Theodore M. Schad, and George William Sherk. (2001) "Water policy in the United States: A perspective," *Water Policy*, vol. 3, p188

14. Collier, Sam. (2006) "Regional institutions for managing water resources," Paper No. 2006-004, Coastal Rivers Water Planning and Policy Center, Georgia Southern University. Statesboro, GA, p6

15. ASCE, op. cit., note 6, p65

16. Braswell, Staci, Florida Farm Bureau Federation, January 20, 2011 letter to Representative Trudi Williams Chair, Florida House of Representatives Select Committee on Water Policy, 2 pp

17. "Senate committee refuses to confirm Swiftmud chief." (2008) *St. Petersburg Times*, 27 April

18. Gonzalez, George A. (2005) "The comprehensive Everglades restoration plan: Environmental or economic sustainability?" *Polity*, vol. 37, no. 4, pp466–490

19. Florida Commission on Ethics. (2010) "Memorandum of voting conflict for county, municipal, and other local public officers (Form 8B)," www.ethics.state.fl.us/, accessed 13 January 2010

20. Oral History Program. (2005) Estus Whitfield, 29 August

21. Foran, Paul G., and Janice A. Beecher. (1995) Survey of eastern water law: A report to the Illinois Department of Natural Resources, www.docstoc.com/docs/20816412/Survey-of-Eastern-Water-Law, accessed 10 January 2011, p63

22. Florida Department of Management Services. (2009) *Annual Workforce Report 2009–20010*, Florida Department of Management Services, Tallahassee, FL

23. Caputo, Marc and Steve Bousquet, "Gov. Rick Scott Unveils Budget of Deep Cuts to Spending, Taxes" *St. Petersburg Times*, February 8, 2011

24. Scott, Rick "Let's Get to Work, the 7-7-7 Plan" February 7, 2011

25. *Palm Beach Post*, "It's Florida, Not Washington: Scott's Budgetary Focus Should be This State" February 8, 2011. *Sun Sentinel* Editorial Board, "Too Much of Gov. Scott's Budget Stretches from Unrealistic to Draconian." February 9, 2011. News-Press, "Lawmakers Must Stand Up to Scott." February 9, 2011. *Florida Today*, "Our View: Gov. Scott's Budget. Spike the tax cuts and find most responsible ways to reduce spending." February 9, 2011. *Miami Herald*: "Our Opinion: Scott's Budget of Extremes a Nonstarter" February 9, 2011

26. Klein, Christine A. (2007) "Water transfers: The case against transbasin diversions in the eastern states," *UCLA Journal of Environmental Law and Policy*, vol. 25

27. Viessman, Warren, Jr. (2006) *State Water Resources Planning in the United States*, ASCE Press, Reston, VA, p10

28. Oral History Program. (2001) John DeGrove, 1 December

29. In 2009, the Florida Department of Environmental Protection issued 19,913 water-related permits plus 9,282 in other FDEP programs. "FDEP statistical abstract," www.dep.state.fl.us/secretary/stats/permits.htm, accessed 10 July 2010

30. Dzurik, Andrew. (2002) *Water Resources Planning* (3rd ed), Rowman & Littlefield, Lanham, MD, p110

31. Bilenky, William S. (2009) "An alternative strategy for water supply and water resource development in Florida," *Journal of Land Use and Environmental Law*, vol. 25, no. 1

32. Burney, Louis C., Tom Swihart, and Janet G. Llewellyn. (1998) "Water supply planning in Florida," *Florida Water Resources Journal*, October, pp27–28

33. [U.S.] Water Resources Development Act of 2000, Section 601, Comprehensive Everglades Restoration Plan, Section (h)(2)

34. The Model Water Code did not recommend only "general" supervision of the water management districts

35. Oral History Program. (2004) Derrill McAteer, 22 June

36. Oral History Program. (2004) Stanley Hole, 17 December

37. Florida Senate, Committee on Environmental Preservation and Conservation. (2009) *Interim Report 2010-114*, September

38. McAteer. Derrill. (1978) "Third Annual Meeting Summary" Havana, Florida, Northwest Florida Water Management District, Tallahassee, October 22–24, 1978, p39

PART II

"CELEBRATING DIVERSITY":
FIVE PHILOSOPHIES FOR FIVE
WATER MANAGEMENT DISTRICTS

CHAPTER FOUR

Five Philosophies for Five Water Management Districts?

T he five water management districts operate under the same statute but approach water management in five distinctly different ways. The Introduction told the story (almost a parable) about the trip that the governing board of the Suwannee River Water Management District took to south Florida in the early days of the new district. The Suwannee district decided to avoid building and maintaining expensive structures like those in south Florida. The policy for Suwannee, expressed in the Oral History interview of their board chair, was to be: *God made it and gravity runs it.*

That is a pithy and entertaining phrase. It made sense for the Suwannee River Water Management District because it fit their historical, demographic, political, and hydrological circumstances. However, the South Florida Water Management District was entangled in a different history and constrained by different circumstances. Different policies were required, as was recognized by the Suwannee River board chair in the same Oral History interview:

Q. Do you think that − using that example − was that a difference in mentality between your district and theirs, or did they just have a different set of missions?

A. Finlayson: No, it's a difference in geography, to begin with. It started out as the South Florida Flood Control District. They were not using non-structural. They were making structures and pumping water. That's what it took. They're flat. We're hilly. Gravity can't run theirs, because it won't go anywhere. It was a difference in geography, it was a difference in need, and it was a difference in philosophy as well. It was all of that.[1]

John Finlayson was right. The five water management districts have somewhat different philosophies of water management. The boundaries of the districts are based on surface watersheds and vary somewhat in the fractions of the state that are their respective responsibilities. Suwannee, the smallest district, has about 13 percent of the state. The largest district, South Florida, has responsibility for about 30 percent of the state. The differences in budget and staffing are much larger. Although under the "general supervisory authority" of the state, the districts operate largely on their own.

There are some sound hydrologic reasons for the districts' water management differences. There is not one Florida in regard to water resources. It is a big state. From Pensacola at the western edge of Florida to Key West at the tip of the Florida Keys is more than 835 road miles. Several climate zones stretch down the peninsula. Large springs are sprinkled across many Florida watersheds but are absent from others. The Floridan Aquifer is one of the largest in the world and underlies much of the state, but other aquifers are more significant in some parts of the state. West Florida is marked by a number of large and small rivers that originate in Alabama and Georgia, but other sections of the state are dominated by lakes and vast wetlands. In south Florida, the rivers are fewer and smaller, except the metaphorical "River of Grass" that Marjory Stoneman Douglas devised for the Everglades.

These hydrologic and environmental differences interact with differences in how Floridians use the water resources. Only three of the districts (Northwest Florida, Suwannee, and St. Johns River) must deal with the complications of sharing watersheds or aquifers with other states. Four of the districts must consider the special needs of springs, but the South Florida Water Management District has no remaining springs. Although all of the districts have significant wetlands, only the South Florida Water Management District has the Everglades. Agricultural irrigation is a major user of water in some water management districts, but not in all. Geography, regional water management needs, and different ideas in the different regions result in different approaches to water management in north Florida and south Florida. What might work in south Florida might fail in north Florida, and vice versa.

Differences in water management philosophy have been developing for almost 40 years since the creation of the Water Resources Act of 1972. Since then, the districts have had diverse responsibilities, ranging from water use permitting to water shortage management to flood plain management to flood control (in the case of two of the districts). The modern environmental movement, beginning in that same era, prompted the districts to undertake many water resource restoration projects, attempting to undo damage created in the previous century of state development. In 1982, the state gave the districts additional responsibility for regulating the construction of wells and stormwater runoff, which moved the districts more firmly into concerns about water quality. In 1995, the state directed the districts to undertake "environmental resource permitting," which combined the old state dredge-and-fill permitting program with stormwater permitting. In 1997, the state again expanded the districts' responsibilities by requiring them to develop regional water supply plans. In 2005, the legislature gave them major new responsibilities for assisting local governments in developing alternative water supplies. At all times, the districts depend upon public support to carry out their large responsibilities.

Today, some differences among the districts are based on regional characteristics such as climate, hydrology, topography, biology, demographics, and water use patterns. Sometimes, however, the districts are different simply because of historical accidents. Are there five different ways to manage Florida's unique array of water resources?

NOTE

1. Oral History Program, op. cit., note 1

CHAPTER FIVE

Suwannee River: God Made It and Gravity Runs It

*T*he Suwannee River is the defining symbol of a part of Florida, but the region also sustains many enormous springs, five other major river systems, scores of lakes, and the state's most pristine coastal system. These remarkable resources are under the care of the Suwannee River Water Management District. The modern district is not the state's first attempt to manage the Suwannee River. In 1957, the legislature created the Suwannee River Authority to promote navigation and recreational tourism. The Authority's approach to managing the Suwannee River was different from modern preferences. For example, in the late 1960s, the old Authority strongly opposed designation of the Suwannee as a national Wild and Scenic River.[1] They even built a crude dam at Suwannee Springs in an effort to assure year-round navigation. The member of the old Authority who drafted the resolution opposing the Wild and Scenic River designation later served as chair of the governing board of the new water management district.[2]

In the 1976 statewide constitutional referendum on property taxes for the water management districts, the voters in the Suwannee District region voted against the proposition by a margin of 64 percent to 36 percent. The statewide referendum nonetheless authorized a property tax for each of the five districts, including Suwannee River. Local politicians still inveigh against high taxes, big government, and onerous regulation. Nonetheless, the Suwannee River Water Management District has responsibility for the water resources of about 7,640 square miles of Florida. The watershed's 320,000 residents are only about 1.5 percent of the state's population.

The Suwannee, the namesake of the water management district, is the second largest river in Florida in terms of flow. The Florida Department of Environmental Protection designated it an "Outstanding Florida Water" in 1979, which afforded it special water quality protection. Although the effort in the 1970s and 1980s to name it a National Wild and Scenic River failed due to local opposition, the Suwannee River strongly merits the designation. It is one of the few major rivers in the southeast that has escaped dam-building. It was a close call for the basin because the U.S. Study Commission on Water Resources recommended a number of major dams in the Suwannee watershed in 1963.[3] Although the Suwannee is free of dams today, it is still threatened in other significant ways. The region is attracting more residents as it is discovered by new migrants to Florida and by relocation within the state. Agricultural practices in the basin are harming water quality in both surface and ground waters.

The main channel of the Suwannee is not the only remarkable water resource in the district. Wetlands occupy a fifth of the district's land area. There are also several large and relatively unpolluted Suwannee River tributaries such as the Withlacoochee (the northern one of two rivers with the same name), the Santa Fe, the Ichetucknee, and Alapaha Rivers. The Alapaha is one of the state's unusual "disappearing" rivers. About two-fifths of the time, its entire flow is captured by sinkholes in the riverbed. After going underground, the river flows southward almost 20 miles before it emerges in Alapaha Rise Spring. The Santa Fe, Steinhatchee, and Aucilla Rivers are other streams in the district that dive underground for part of their courses.

The combined discharge of springs in the watershed is immense, totaling more than 7,200 cubic feet per second (4.65 billion gallons per day).[4] More than 250 separate springs have been identified in the Suwannee watershed, which has more large springs than any other place on earth. Twenty-one springs discharge more than 100 cubic feet of water per second (making them "first magnitude" springs). Large springs like Manatee, Ichetucknee, Troy, and Ginnie are nationally famous and major recreational resources.

The geographic area within the district includes watersheds outside that of the Suwannee River. The Steinhatchee, Aucilla, Wacissa, Waccasassa, Fenholloway, and Econfina Rivers also flow to the Gulf of Mexico. The watershed management task for the district is complicated by the fact that about half of the watershed is in Georgia, including the origins of the Suwannee River in the Okefenokee Swamp (a National Wildlife Refuge).

The Suwannee River is, in a sense, three different rivers. The upper river, from its origin in Georgia's Okefenokee Swamp to the town of White Springs in Florida, is relatively small and narrow. The water of the upper river is naturally tea-colored from swamp drainage. For a Florida river, its banks are remarkably high and make canoeing or kayaking an unusual experience. When the river is low, paddles noisily scrape the limestone bottom of the upper river. (That limestone is the same rock making up the Floridan Aquifer.) At high flow, recreational paddlers can shoot Big Shoals, Florida's only Class III river rapids. Many paddlers have upset their canoes here, including the author and his wife.

The middle river, from White Springs to Wilcox (36 miles from the Gulf of Mexico), is the second version of the Suwannee River. Here it is closely connected to the Floridan Aquifer and is marked by a combination of large springs that contribute large volumes of clear water to the river. At low flows, the clear water forcing its way out of the earth makes up more than half of the river flow. Nitrogen concentrations in the river increase because a larger fraction of river flow comes from groundwater, which has artificially high levels of nitrogen from human activities. The clear water of this section of the river is perfect for passing sunshine through the transparent depths. Unfortunately, the transparent water and abnormal concentrations of nitrogen combine to grow dense populations of nitrogen-fertilized filamentous algae.

When the river is at flood stage, some major springs in this section go into reverse. Rather than discharging to the river as springs normally do, a portion of the Suwannee River flows into the spring openings. This reversing of flow reduces flood peaks to some degree. The flows stored in the aquifer are released into the river later when flood stages end.[5] That cyclic process tends to even out the annual flows of the river.

In the lower third of the Suwannee, which empties into the Gulf of Mexico, a different kind of river appears. The river is now large and wide, with average flows of more than 10,000 cubic feet per second. There are no tributaries of any significance in this section. This part of the river includes a state Aquatic Preserve and a 53,000-acre National Wildlife Refuge, with large areas of salt marshes, tidal creeks, and an active oyster and clam industry. Gulf of Mexico tides exert their salty influence many miles upstream. The Suwannee estuary, and the coastlines on either side of it, are among the least developed and least polluted coastal regions in the United States. The Suwannee contributes 60 percent of the fresh water flowing out into the state's relatively pristine Big Bend coastal region, and is the largest commercial oyster harvesting area in Florida, aside from Apalachicola Bay.[6]

Even though the many springs on the river tend to even out the flow, the Suwannee watershed does experience both very low and very high flows. The drought of 1998–2000 was exceptional, but the drought of 2006–2007 was even more severe and set new records in the Suwannee River District. For example, the Santa Fe River had the lowest flows ever recorded for the months of July, August, and September in 2006. The watershed also experiences bouts of severe flooding. In 1948, spring floods inundated 500 square miles of the Suwannee River system. Major floods occurred in 1948, 1959, 1964, 1973, 1984, 1986, 1991, 1998, and 2009. New flood records were set in April 2009 on part of the system. The (northern) Withlacoochee River, a large tributary to the Suwannee, crested 2.65 feet above the previous record set in 1948. Downstream on the Suwannee River, flooding occurred but was alleviated by the floodwaters being able to move freely into available groundwater storage due to the preceding dry season.

The Suwannee River Water Management District recognized early in its existence that flooding was a major concern. A first step in response was to map the floodplain of the river. The mapping was a key element of the district's philosophy to leave the river alone and to rely upon nonstructural means of watershed management. The mapping helped local governments exercise their land use powers to keep development out of the way of damaging floods. However, not all of the counties adopted adequate regulations to keep homes and other development out of harm's way during flooding. The district, in its own permitting programs, strongly discourages development in riverine floodplains.

Calls for channelization and flood control levees after floods have been modest because the damage from high water levels has been minimized.

One of the Suwannee District's most successful programs has been the use of state dollars to buy land, mostly floodplains. Don Morgan, the district's second executive director, was one of the original members of the 1971 task force that formulated the resource management recommendations enacted by the legislature in 1972. He came from the Central and Southern Florida Flood Control District and helped the Suwannee District decide on a nonstructural path. Earl Starnes, a long-time Suwannee board member, drew on his previous experience as a Dade County commissioner and head of the Florida Division of State Planning. He described a decision, along with Don Morgan, to "buy the Suwannee River":

> Don and I one day were talking about this thing, and Don said, why do we not just go ahead and buy the Suwannee River? I said, well, I think that is a good goal. We figured, then, it would only cost $100,000,000 to buy the whole river, so what we did, the board appointed me as chair of the acquisition-planning committee[. . .]What we basically decided to do was to go ahead and take the river in the three sections, the northern, middle, and southern Suwannee, and have staff develop a plan around those areas, an acquisition plan which would identify lands that should be acquired. In the process, we did not know where the floodplain was, so one of our early projects was to go ahead and survey the floodplain, because we did not know where the 100-year floodplain was.

> So, we had the Suwannee River Basin flown and surveyed by air, and then we had survey crews on the ground and they established the benchmarks. Then, we knew where the 100-year floodplain [was]. Our decision was then to buy everything in the 100-year floodplain that was available. That process is still ongoing. Today, the district is the largest land-owner on the river.[7]

In fact, the district owns an extraordinary fraction of its rivers. It has title to, for example, 330 miles of river frontage. For the Suwannee alone, the district owns 82 miles of the upper river frontage, 29 miles of the middle river, and 31 river miles of the lower river. It also owns 50 miles of river frontage of major tributaries of the Suwannee River, such as the Alapaha, Santa Fe, and Withlacoochee Rivers. The district owns 63 percent of the length of the Econfina River and major fractions of the frontage along the Steinhatchee and Wacissa Rivers. They also have acquired a number of the large springs in the basin, as well as large wetland areas, such as Malloy Swamp, Hixtown Swamp, California Swamp, and Waccasassa Flats. Don Morgan envisioned a day a century in the future when the district's acquisition of the Suwannee's floodplain would be as valuable for a densely populated north Florida as Central Park was for New York City.[8] Most of the money for buying land came from state land acquisition programs like Save Our Rivers and Florida Forever because the property tax base of the Suwannee District is relatively small and property taxes are unpopular.

From time to time, downstate interests propose diverting water from the Suwannee watershed to meet their growing demands. The district strongly opposes any such removal of water from the Suwannee. Although any part of the riverine system could be damaged by long-distance transportation of water out of the area, effects on the Suwannee estuary from diversion might be the most severe. The U.S. Geological Survey pointed to a wide range of possible effects from lowering flows of fresh water into the Suwannee estuary, including:

- Increased salinity levels in the estuary
- Reduced floodplain habitat
- Changes in forest community types
- Decline of tidal forests and transition into brackish marshes
- Decreased species richness
- Loss of wildlife habitat
- Loss of soils due to oxidation during times of drought
- Increased vulnerability to fire[9]

The district still has a policy of nonstructural floodplain management and does not operate any dike, levee, or dam. That policy has protected the Suwannee from the harmful structural alterations so common elsewhere in Florida and other places. The slogan "Gravity runs it" made sense for a new agency, with limited budget, charged with the responsibility for maintaining a watershed basically in good shape and with few challenges.

That old slogan has a shortcoming, however. The time is past when any water management district could mostly watch gravity do its work. New challenges to sustainability are present even for lightly settled parts of Florida, and modern water management requires substantial resources and active attention. As staff at the district sometimes say, it is time for the district to "grow up and be a real water management district."

IMPROVING WATER QUALITY

At present, water quality in the region is generally good. The major exception is excessive levels of nitrogen, which was detected at elevated levels as early as 1985. Natural groundwater concentrations were less than 0.2 mg/L in the 1950s and 1960s. At some monitoring locations today, they have reached 5 mg/L.[10] The middle Suwannee basin, which is only 9 percent of the watershed area, contributes 46 percent of the nitrogen flowing into the estuary at the Gulf of Mexico.[11]

The primary source of the nitrogen is agriculture, both from cropland fertilization and from confined animal facilities like dairies and chicken farms. It can take decades for the nitrogen from farms to be released in spring vents. The nitrogen load from milk cows, beef cattle, and poultry in the basin is estimated to be about 16,600 tons per year, almost seven times that from humans.[12] Row crops, poultry operations, and dairies are estimated to be the source of two-thirds of the nitrogen pollution placed into the region's groundwater.[13] It is essential to control the nitrogen pollution of the aquifer now to prevent it from further polluting the springs and rivers decades in the future.

The district recognizes that excessive nutrients from farms, homes, and small cities are very significant resource management issues. Until recently, however, the Suwannee District opposed the state's efforts to establish nutrient loading limits for the river in the Total Maximum Daily Load program. A large part of that opposition was grounded in protecting the interests of agriculture, the largest polluter in the district.

REFORMING AGRICULTURAL PRACTICES

Industrial agriculture is relatively new in the Suwannee River watershed. Irrigated acreage has increased dramatically in the last few decades, as well as other more-intense forms of farming. There were 46 dairy operations in 2009, housing more than 34,000 cows. More than 119 poultry operations in the area had more than 30 million birds that year.[14]

As explained later in Chapter 13, Farming the Water, some of the dairies in the Suwannee River basin moved there from south Florida after their pollutant impact became intolerable in their original locations. The dairies were paid to move out of the Lake Okeechobee basin in the 1980s but were welcomed in the Suwannee River region. This amounted to short-sightedly moving a pollutant problem from one part of the state to another.

Dairies are not the only source of nitrogen contamination. Another is associated with pet horses. There are 500,000 horses in the state, mostly in central and south Florida.[15] North Florida grows 700,000 tons of hay a year, much of which is fed to horses. The Institute of Food and Agricultural Sciences at the University of Florida recommends applying 500 pounds of fertilizer per year to each acre of hay production, thereby adding more nitrogen to the troubled Suwannee region aquifer system.[16]

By the 1990s, water pollutants reached a level that was hard to ignore. State regulation seemed inevitable. The water management district sided with farmers in opposing direct regulation, favoring instead the voluntary Suwannee River Partnership. The Partnership relies on voluntary participation by farmers, aided by financial assistance programs from the district, state and federal Departments of Agriculture, and the Florida Department of Environmental Protection. The U.S. Department of Agriculture alone has spent more than $25 million on projects on farms to reduce pollution.[17] (This direct expenditure is in addition to the general agricultural subsidies described in Chapter 13.) The great majority of dairies and poultry operations have nutrient management plans (as also do 125,000 acres of cropland in the watershed).[18]

Despite paying farmers many millions of dollars to control their pollution and avoid direct regulation, the Partnership has seen only limited success:

> The "Suwannee River Partnership" is intended to help farmers reduce their impacts on water resources, especially the loading of nitrogen to the river and aquifers, while also promoting profitable operations. The Partnership tries to avoid regulation by providing financial assistance from the district, as

well as state and federal agencies, in implementing Best Management Practices. Since being formed in 1999, the vast majority of poultry, dairy, and crop farms have agreed to implement the BMPs. However, the Partnership has seen only partial success in water quality trends. In some places, nitrogen has actually increased, while others have seen some decreases and others no apparent change at all.[19]

Agricultural operations in the Suwannee River Water Management District are as powerful, and are treated as gingerly with regard to water regulation, as anywhere in the state. In 2011, the present and previous chairs of the governing board operated farms in the district. Nevertheless, it is increasingly clear that additional district regulation will be necessary to address this biggest water quality problem.

BUILDING FISCAL SOLVENCY

The tax base of the Suwannee River Water Management District is the smallest of all of the five water management districts. It has only 65 employees – no more than it had a decade ago before problems in the watershed became more serious. Still, the Suwannee District raises only about 8 percent of its budget from property taxes, partly because it assesses a lower millage rate than state law allows. The legislature made the district's budget problems even worse in 2008 by directing all of the water management districts to reduce their millage rates by 10 percent. After assessing a millage rate of 0.49 mills for 14 years, the governing board of the Suwannee District reduced the millage that year to 0.44 mills. The lower millage rate, and much-reduced state financial assistance, created a fiscal crisis for the Suwannee District. The district was forced to suspend part of its water quality monitoring and also its work on establishing minimum flows and levels. In 2010, the governing board voted to slightly increase the tax level to raise about $750,000 to fund the minimum flows and levels effort. Under pressure to not increase tax rates at all during a recession, the board quickly rescinded the vote and reduced other programs to provide at least some funding for minimum flows and levels.

Part of the fiscal problem is self-inflicted. The Suwannee District follows the practice of other water management districts in subsidizing the cost of handling permit applications. The district recovers only 16 percent of the cost of processing consumptive use permits, and only 21 percent of the cost of environmental resource permits.[20] That subsidy provides an incentive for water use and land development.

A possible alternative source of funding for the Suwannee River Water Management District, as well as for the other four water management districts, would be a state water use fee. If a water use fee substituted, in whole or part, for the property tax, a large and stable revenue source would be available for critical water management functions. This option is described in Chapter 12, Charge a Few Nickels for Water.

PROTECTING NATURAL FLOWS

In popular imagination, the biggest threat to the Suwannee is water pirates from south Florida who want to run a pipe to the river and suck out much of the water. Watershed residents have long memories for such talk, which has been floated over the years. For example, a 1997 study done for the West Coast Regional Water Supply Authority (predecessor to Tampa Bay Water) concluded that it was feasible to pipe water to Tampa from the Suwannee River. The U.S. Army Corps of Engineers released a study with similar conclusions that same year and even included a map of a possible pipeline route reaching from the Suwannee to as far south as Sarasota.[21] More recently, in 2003 the Florida Council of 100 (an organization of big businesses) published a report commending the possible advantages of a water transfer. Lee Arnold, the chair of the Council of 100, called the Suwannee River "the Saudi Arabia of water."[22] The proposal was severely criticized in the news media. Half of the state's 67 counties passed resolutions condemning the proposal.

The idea is not dead, however. In 2008, the Florida section of the American Water Works Association proposed investigation of a statewide water supply "grid." This was considered at the 2008 Water Congress meeting but received very little support and was not part of the Congress's final recommendations.

The concern about a pipeline is understandable but may be misplaced. Commissioner of Agriculture Adam Putnam opposes such a transfer, saying that it would put the state "on the brink of a civil war."[23] It is many scores of miles from the Suwannee River to any urban area, which would make a pipeline expensive and difficult to build. The energy costs of pumping water though such a long pipeline would also be large, perhaps making it infeasible from engineering and economic perspectives. A more realistic threat to natural flows is new withdrawals in the Suwannee watershed itself or in bordering regions. Either could take so much water out of the aquifer that it would lower groundwater levels in the Suwannee watershed. The Suwannee River Water Management District calls itself a "victim of groundwater pumping" located within its own district, in Georgia, and in the adjoining St. Johns River Water Management District.[24] Pumping in the Jacksonville area and southeast Georgia has moved the groundwater divide between those areas and the Suwannee River; more than 2,000 square miles of aquifer that once flowed into the Suwannee watershed now flows out of it to the east.[25] The two districts are conducting studies of the aquifers of northeast Florida, which already have lowered spring flows. Already, it is known that excessive groundwater withdrawals outside the Suwannee area have caused dramatic changes.[26]

Both the St. Johns and Suwannee River Water Management Districts will face significant challenges. This is one of the reasons it is critically important that the Suwannee district adopt protective minimum flows and levels. A shrinking budget makes that essential task much harder to accomplish.

TAKING ON HARD ISSUES

The water problems of the Suwannee watershed are growing rapidly because the area is reaching the threshold of growth that other regions of Florida did long ago. The Suwannee River Water Management District has not yet had to tackle many hard problems because few have presented themselves. Low numbers of people and relatively small-scale farming have not yet irreparably damaged the Suwannee. Yet it is not clear that the district is prepared for a future of more serious problems ahead.

For example, the Suwannee District does not require that all agricultural water use be accurately recorded and then reported regularly to the district for monitoring. Unlike some other districts, the Suwannee routinely issues water use permits for 20-year periods, which locks-in water use with few opportunities for modification if future water problems occur. The district does not have a single designated water planning position to help it prepare for its water future. The Suwannee District will soon have to take on hard issues that are not as popular with residents and taxpayers as buying up riverfront land with state money.

NOTES

1. "Fuqua asks that Suwannee be dropped." (1968) *St. Petersburg Times*, 31 August

2. Oral History Program. (2004) J. C. Camp, 5 October

3. U.S. Study Commission, Southeast River Basins. (1963) *Plan for Development of the Land and Water Resources of the Southeast River Basins*, U.S. Study Commission, Southeast River Basins, Atlanta, GA, pp4–6

4. Florida Department of Environmental Protection, Division of Water Resource Management. (2001) *Basin Status Report: Suwannee*, p36

5. Giese, G. L., and M. A. Franklin. (1996) *Magnitude and Frequency of Floods in the Suwannee River Water Management District*. U.S. Geological Survey WRI Report 96-4176, p9

6. Mattson, Robert A. (2002) "A resource-based framework for establishing freshwater inflow requirements for the Suwannee River estuary," *Estuaries*, vol. 25, no. 6B, pp1333–1343

7. Oral History Program. (2000) Earl Maxwell Starnes, 17 June

8. Personal communication with Kirk Webster, SRWMD, October 29, 2009

9. Darst, Melanie R., Helen M. Light, and Lori J. Lewis. (2002) *Ground-Cover Vegetation in Wetland Forests of the Lower Suwannee River Floodplain, Florida, and Potential Impacts of Flow Reductions*, U.S. Geological Survey WRI Report 02-4027, p27

10. U.S. Geological Survey. (2004) *Suwannee River Basin and Estuary Initiative: Executive Summary*, Open File Report 2004-1198

11. Dedekorkut, Aysin. (2005) "Suwannee River Partnership: Representation instead of regulation," in John T. Scholz and Bruce Stiftel, *Adaptive Governance and Water Conflict* (2005) Resources for the Future, Washington, DC, p27

12. Hallas, John F., and Wayne Magley. (2008) *Nutrient and Dissolved Oxygen TMDL for the Suwannee River, Santa Fe River, Manatee Springs (3422R), Fanning Springs (3422S), Branford Spring (3422J), Ruth Spring (3422L), Troy Spring (3422T), Royal Spring (3422U), and Falmouth Spring (3422Z)*, Florida Department of Environmental Protection, Bureau of Watershed Management, Tallahassee, FL, p47

13. Florida Department of Environmental Protection, op. cit., note 4, p80

14. These estimates of farm animal populations in the Suwannee come from an October 29, 2009 conversation with Kirk Webster, deputy executive director of the Suwannee River Water Management District. The estimates are substantially lower than other recent estimates, such as those by Darrell Smith, Holly Stalvey, and Charles T. Woods, Jr. in "Protecting North Florida Springs," *Florida Watershed Journal*, vol. 1, no. 3, Fall 2008, p12. The numbers of dairy cows and chickens in the watershed vary somewhat but appears to be cyclic

15. Chambliss, C. G., E. L. Johnson, and I. V. Ezenwa. (2006) "Pastures and forage crops for horses," *Document SS-AGR-65*, Florida Cooperative Extension Service, Institute of Food and Agricultural Sciences, University of Florida, Gainesville, FL

16. Mislevy, Paul. (2010) "Horse hay: A cash crop for Florida cattlemen," Range Cattle Research and Education Center, Agronomy Department, and Dr. E. L. Johnson, Animal Sciences Department, UF/IFAS, http://rcrec-ona.ifas.ufl.edu/in-focus/IF7-7-06.shtml, accessed 3 July 2010

17. Cooperative Conservation America. (2009) "Case study: Suwannee River Partnership," http://cooperativeconservation.org/viewproject.asp?pid = 783, accessed 19 December 2009

18. Suwannee River Water Management District. (2008) *Comprehensive Annual Financial Report for the Year Ended September 30, 2008*. Suwannee River Water Management District, Live Oak, FL

19. Suwannee River Partnership. (2007) *Partnership News*, vol. 4, no. 7

20. Florida Senate, Committee on Environmental Preservation and Conservation. (2008) *Agency Sunset Review of the Water Management Districts*, Report Number 2009-207

21. Geraghty & Miller. (1977) *Hydrologic and Engineering Evaluation, Water Resources Management Study of the Four River Basins Area, West-Central Florida*, vol. I, sections 1 through 7, prepared for the U.S. Army Corps of Engineers, Jacksonville District, Jacksonville, FL

22. Barnett, Cynthia. *Mirage: Florida and the Vanishing Water of the Eastern U.S.*, University of Michigan Press, Ann Arbor, MI, p111

23. Bousquet, Steve "For Putnam, it's all about the water" *St. Petersburg Times*, January 19, 2010.

24. David Still, Executive Director, SRWMD, letter to [Florida] Governor Charlie Crist, March 5, 2010

25. Grubbs, J. W., and C. A. Crandall. (2007) *Exchanges of Water between the Upper Floridan Aquifer and the Lower Suwannee and Lower Santa Fe rivers, Florida*, U.S. Geological Survey Professional Paper 1656-C

26. Suwannee River Water Management District, "Water Supply Assessment, 2010" SRWMD, Live Oak, Florida, 109 pp

CHAPTER SIX

Northwest Florida:
The Forgotten District

Northwest Florida legislators hobbled this district's effectiveness in 1976. As discussed in Chapter 2, the legislature that year authorized a constitutional referendum to grant the five water management districts a property tax levy of up to one mill ($1.00 per $1,000 of assessed property value). That small tax would provide a revenue foundation for effective water management throughout the state. Combined with supplemental state or federal appropriations, the property tax would give the water management districts the ability to take on difficult and expensive water management tasks.

Except the Northwest Florida Water Management District. Legislators in northwest Florida managed to put into the referendum language a provision that the millage rate cap for the Northwest Florida Water Management District would be 0.05 mill, only one-twentieth of the 1.0 mill maximum granted to the other four water management districts. Don Duden, later the assistant executive director of the Florida Department of Natural Resources, recalled how the speaker of the Florida senate felt about the proposed tax for water management:

> He [Dempsey Barron, from Bay County in northwest Florida] was total and absolutely hell bent there was not going to be a water management district in north Florida when it all started. Then he realized the error in that judgment and decided yes, there will be, but you are not going to tax my people to do it. He was pretty vocal; there was just no doubt – I don't think in anybody's mind – that there wasn't going to be any tax out there. That was a dumb decision because how are they going to operate[?] They got to do something, so he ultimately agreed to that cap. That was all that he would agree to.[1]

The lower millage rate for the Northwest Florida Water Management District appeared on the ballot, along with the higher maximum rate for the four other water management districts. Voters statewide approved it. As a consequence of the legislative maneuvering, Northwest Florida lacks the financial base for building water resource sustainability that the other districts possess. It depends mostly on annual appropriations from the state legislature. In effect, citizens in the rest of Florida tax themselves for water management services in the Panhandle.

This financial handicap restricts the water management abilities of the Northwest District and is one reason it is often overlooked or forgotten. The district is also little-noticed because the Panhandle is remote from the urbanized central and southern parts of the state, which are the primary concerns of statewide news media coverage. The region also suffers from the lack of a unified water identity because a wide span of north–south rivers cuts through the district, unlike the single rivers that define the Suwannee and St. Johns River Water Management Districts. About the only part of the district that receives wide notice is the interstate contest over the Apalachicola River, which is discussed later in this chapter.

It is not as if the 1.4 million people of northwest Florida do not have water resources that merit protection and management. In fact, the 11,300 square miles of the Northwest District have some of the most extraordinary water resources in the state, combined with one of the state's highest rates of population growth. The northwest region is biologically distinctive and one of the five richest in biodiversity in North America. Many animals and plants from northern regions of the continent reach their southern range limit in the Florida Panhandle, while many peninsular species reach the region as their northern limit. The combined effect is high biological diversity. No area of comparable size in the United States has more species of frogs or snakes. Apalachicola Bay may be the most biologically productive estuary in the Northern Hemisphere.[2]

The Northwest District has three of the five largest rivers in the state: Escambia, Choctawhatchee, and Apalachicola. These three rivers have more rare fish species than any other in Florida.[3] The rivers of the region are some of the cleanest in the southeastern United States. They have not undergone much in the way of impoundment or channelization, at least in the parts of their watersheds within Florida. The district has no flood control dams (although there are some reservoirs for water supply and other purposes).

Like the Suwannee District, major fractions of the agency's watersheds are in the neighboring states of Alabama and Georgia. The Apalachicola, the largest river in Florida, has its source in the Flint and Chattahoochee Rivers. These two Apalachicola River tributaries merge in a U.S. Army Corps of Engineers impoundment at the state line (Lake Seminole). The large rivers of the District help define the region but also create large estuaries like Choctawhatchee, St. Andrews, Pensacola Bay, and the massive Apalachicola Bay system.

The district counts five "first magnitude" springs (Wakulla, Spring Creek Group, St. Marks River Rise, Gainer Spring Group, and Jackson Blue Spring), as well as three dozen second magnitude springs. Wakulla Springs, now within a state park, is thought to have the largest spring vent in North America, measuring 50 feet by 82 feet. Wakulla appears to have a remarkably direct hydrological connection to Spring Creek, located 10 miles to the southwest on the coast of the

Gulf of Mexico. When Spring Creek temporarily stopped flowing during the drought of 2006, the flow of Wakulla Spring doubled. Hydrologists theorize that perhaps the drought had lowered groundwater levels so much that saltwater blocked the vents at Spring Creek and aquifer water instead flowed out of the vent of Wakulla Spring.[4]

Some of the springs in the district have artificially high amounts of nitrogen. Jackson Blue Spring near Marianna, for example, saw increases from 0.3 mg/L in 1960 to 3.5 mg/L in 2009. The district compares that amount of nitrogen to "a free hot dog in every glass" because the nitrites in an average hot dog is about the same as in a 16-ounce glass of water from the spring.[5]

The region has many lakes but not as many as central Florida. Lake Jackson, near the state capital of Tallahassee, is one of Florida's notable "disappearing" lakes. It cycles, usually at intervals of decades, between being at its usual level and having rapidly lost almost all of its water. In the most recent disappearance in 1999, most of the 4,000 acres of lake water drained into the bottom of Porter Hole Sink in just a few days. The rest of the lake drained in the spring of 2000 into another sink. The lake has no outlet and no river or spring flowing into it, so the lake coming back is completely dependent on rainfall. After going dry in 1999, Lake Jackson did not rebound for more than a decade.

Like the lakes, the rivers cycle naturally back and forth from low to high flows. Several rivers in northwest Florida broke low-flow records in 2006. At the other extreme, the major rivers flood periodically and inundate their floodplains. The damage from river floods is less than it might be otherwise because intense development has not occurred along their banks. In part, that flood risk has been averted by the water management district buying major fractions of river floodplains.

Historically, land and water were regarded in northwest Florida as a natural resource to be used for as much near-term profit as possible. Longleaf pine forests once stretched from Virginia to Texas and covered more than 10,000 square miles of north Florida. Lacking much else for economic development, the pine and hardwood forests of the region were mowed down in the late nineteenth and early twentieth centuries. Industrial lumber operations scalped almost all of the longleaf pine forests and fed 6.5 billion feet of the state's virgin longleaf pine into lumber mills by the 1930s.[6] Entrepreneurs stepped into the cutover zone and bought the nearly worthless land at very low prices in the Great Depression. The original removal of the virgin pine forests was followed by industrial forestry. The St. Joe Paper Company became the state's largest landowner, holding title to more than 1 million acres, mostly in northwest Florida. The company's philosophy for 60 years was to hold on to the land and grow trees on it. In the late 1990s, the St. Joe Paper Company closed its mill, shortened its name to the St. Joe Company, and became the state's largest real estate company, with more than 406,000 acres within 15 miles of the fast-growing coastline of northwest Florida. The company's goal is "making places" and to "bring high-quality, strategic growth to the last, best part of Florida."[7] This overlooks the fact that north Florida is already a place and that growth may not be the best thing that happens to the quality of life in this region.[8] It is cause for concern that newly inaugurated governor Rick Scott appointed in early 2011 a former vice-president of strategic planning for the St. Joe Company as the new Secretary of the state Department of Community Affairs.[9]

WATER SUPPLY IN THE PANHANDLE

Unlike peninsular Florida, water supply is not a defining issue for the Northwest District. For example, the district has never had to issue a water shortage order. It does not even restrict the number of days per week that lawns may be irrigated. There are some water supply problems, however, caused mostly by the region's urban settlements being located directly on the coast of the Gulf of Mexico. Except for the state capital of Tallahassee, all of the largest cities in the district are located on the coast. All of the public water supply for these cities comes from aquifers, except for Panama City's Deer Point Lake Reservoir. The coastal location of cities makes municipal wellfields vulnerable to saltwater intrusion as they pump more groundwater to serve more residents. For example, in 2008 overpumping of the aquifer for Ft. Walton Beach drove its level down 110 feet below sea level, threatening saltwater intrusion.[10]

The Northwest Florida Water Management District encourages utilities to move their wellfields inland to reduce the threat of saltwater intrusion. Part of the encouragement is very tangible: millions of dollars from the district's portion of the state Water Protection and Sustainability Trust Fund for the development of "alternative" water supplies. The district, although handicapped with the very restrictive constitutional cap on its tax rate, nonetheless has befriended local governments by acquiring and then passing on state and federal dollars. From 2003 to 2008, the district passed on more than $20 million in Florida Forever appropriations to local interests in what it called district "capital improvement grants." Local governments are happy to accept the funds. Only the Northwest District treats shifting a well inland as an "alternative" water supply. It also uses funds from the state Water Protection and Sustainability Trust Fund to help pay for drinking water pipelines and treatment plants.[11]

In 2001 the district developed a regional water supply plan for the fast-growing coastal sections of Santa Rosa, Okaloosa, and Walton Counties, and in 2006 a plan was published for the coastal fringes of Franklin and Gulf Counties. A third plan was prepared for coastal Bay County in 2007. All of these plans were necessary due to fast population growth and the threat that saltwater intrusion posed to public water supply wells close to the shore.

EXTRAORDINARY WATER RESOURCES

Like the Suwannee River Water Management District, the Northwest District has used state dollars for a large and popular land acquisition program. By 2009, the district owned outright, or has conservation easements on, more than 216,000 acres of land. Land prices in much of the Panhandle are lower than in urban parts of Florida, which helped the district in buying large tracts of floodplain and aquifer recharge properties. The 216,000 acres amount to only 3 percent of the geographic area of the district, but they also are 59 percent of the floodplain sought in the district's land acquisition plan. By buying floodplains, the district has permanently prevented development in those vulnerable areas and protected the biological functions of those important systems.[12]

The land purchased by the district often needs environmental restoration, including lands that never recovered from the timbering of longleaf pine forests decades ago. The district has planted millions of longleaf pines on its land and restored thousands of acres in different watersheds. One of their major water restoration activities is assisting state agencies in management of Tates Hell, an enormous expanse of 200,000 acres mostly timbered in the 1960s and 1970s and now part of a state forest. The industrial forestry operators altered the natural hydrology with more than 800 miles of logging roads and ditches to make it more suitable for growing slash pines. Restoring the hydrology and ecosystem functioning of the state forest will require decades. The land acquired by the district with state appropriations by the district is very significant. The holdings contribute directly to water management needs by protecting aquifer recharge areas and preventing development in hazardous floodplains. However, buying land cannot address some of the biggest challenges facing the district.

THE ACF SYSTEM

The biggest water management shortcoming of Florida state government and the Northwest Florida Water Management District in this region has been failure to restrain Georgia's excessive consumption of water from the streams and aquifers that flow into the Apalachicola River. Interstate water management success for this river is necessary because the great majority of the watershed of the Apalachicola River is in Georgia and Alabama. More than 85 percent of the flows in the river come out of the Jim Woodruff Dam where Florida, Georgia, and Alabama meet. State government has the ultimate legal responsibility in negotiating interstate water matters and has been active in the Apalachicola-Chattahoochee-Flint (ACF) River issues. However, the state looks to the regional water management district for both the fundamental science of the river and for shared leadership on this water management issue.

Protection is merited because the Apalachicola River and Bay system is one of the most remarkable on the continent. The Apalachicola River is one of the least polluted rivers in the southeastern United States and supports one of the most productive estuaries in the northern hemisphere. The biological value of the river and bay complex has been recognized by its designation as an International Biosphere Reserve, an Outstanding Florida Water, a national Estuarine Research Reserve, and a state Aquatic Preserve. The river is home to highly diverse flora and fauna, including 131 species of fish.

The Apalachicola has the largest flow of any river in Florida; its headwaters originate hundreds of miles to the north in the foothills of the Appalachian Mountains. Seasonal flooding of the state's largest floodplain is the force that moves essential nutrients into the ecosystem and fuels the oyster industry in Apalachicola Bay, which is a mainstay of the economy of Franklin County. Endangered and threatened species also are at risk. The Gulf sturgeon (*Acipenser oxyrinchus desotoi*) once traveled up many rivers in the northern Gulf of Mexico, but now is limited to very few, including the Apalachicola. Endangered species of mussels that live in the Apalachicola have become a pivot point in interstate water litigation. The threatened Purple Bankclimber (*Elliptoideus sloatianus*) and the fat Threeridge mussels (*Amblema plicata*) are immobile invertebrates and rely on good water quality, flowing water of the right velocity, stable riverbeds, and avoidance of extreme low flows.

Upstream activities threaten all of the desirable characteristics of a naturally functioning river. The harm done to the Apalachicola River comes principally from three related sources: urban overwithdrawals, agricultural overwithdrawals, and structural alterations to the river system. It is their combined effect, rather than each individually, that creates the damage.

Rapidly growing metropolitan Atlanta is the largest urban area in the southeast and is overwhelmingly the location of urban withdrawals in the ACF watershed. The Atlanta Regional Commission projects that water use will increase from 650 mgd in 2001 to 1080 mgd in 2030. Much of that water use is unnecessary because so much of it is used inefficiently. In a report commissioned by the state of Florida, the Pacific Institute found that so much water was wasted that the Atlanta region could easily meet even its 2030 urban demands with the amount of water they use today, by implementing cost-effective water conservation measures.[13] The CEO of the national Alliance for Water Efficiency criticized Georgia's 2008 Water Conservation Plan for having only recommendations that must be "voluntarily adopted by water utilities and local governments."[14]

Nonetheless, a bill approved by the 2010 Georgia Assembly did have some meaningful water conservation provisions. The Georgia Water Stewardship Act prohibits landscape irrigation between 10:00 a.m. and 4:00 p.m. (like Florida policies) and requires the use of the most efficient water-using plumbing devices by 2012 (which exceeds Florida requirements). American Rivers, a national environmental organization, said that "Georgia now leads most states in the nation when it comes to 21st century water supply solutions," and the president of the Georgia Conservancy called it the "most significant, sweeping water conservation policy in Georgia's history."[15]

It is not just urban Atlanta that takes water out of the river system. Agricultural withdrawals in southwest Georgia may do even more harm to the Apalachicola. Downstream, principally in southwest Georgia, there are large and intensive agricultural irrigation operations. In fact, the volume of consumptive use of water in agriculture in the basin far exceeds the consumptive use by cities.[16] In contrast to urban water use in Atlanta, which comes almost entirely from surface water withdrawals, most agricultural water use is supplied from ground water. The Chattahoochee and Flint Rivers lose about six-tenths of a gallon of flow for each gallon pumped out of the aquifer.[17] The amount of agricultural use in the basin is so large that it reduces Apalachicola River flow downstream even more than do urban uses. Agricultural irrigation is a prime factor in converting periods of merely low flow into record drought conditions in southwest Georgia and the Apalachicola River.[18] When river levels are low in a drought, and under the greatest stress, is exactly when farm irrigators pump the most.

Thus, an interstate water agreement on the ACF river system must include effective management of agricultural irrigation, not just municipal withdrawals. This is very challenging, because Georgia and Alabama mostly exempt agricultural irrigation from regulation and even from water use reporting. The Georgia Department of Natural Resources reports that most agricultural water use in Georgia is not monitored and there are only a few studies of overall agricultural water use in the state.[19] Most of the agricultural irrigation in the basin happens in Georgia, which requires no permit for any withdrawal below 100,000 gallons per day. Also exempt from regulation is any withdrawal of less than 150,000 gallons per day "for transportation purposes," any filling of an impoundment, or any farm pond constructed for "fish, wildlife, recreation, or other farm uses." (Section 3, Georgia Environmental Protection Division Rule, 391-3-6-.07, Surface Water Withdrawals). Agricultural irrigation permits in Georgia law have "no term," meaning that they have an unlimited duration and cannot be revoked. (Section 9 of the EPD Rule)

The key point is that Georgia exempts much agricultural water use from the scope of its regulatory authority. This deficiency was noted by the Georgia Water Coalition when it called for Georgia water law to be amended to treat "agricultural permits under the same standards applicable for other uses, including water conservation requirements."[20] Georgia has no effective means of complying with any future controls on agricultural water use. All of the interstate negotiation and litigation have an air of unreality about them if that critical fact is ignored.

Effective water management matters especially in drought years in the Apalachicola watershed because they are drier than average drought years of the past; the driest months of the year are especially diminished. Biological resources have suffered. Along the wide Apalachicola River floodplain, the system has become much drier, which diminishes the vital exchange of nutrients between a river and its floodplain. In effect, the floodplain forest is drying out. Lower flows to Apalachicola Bay also mean higher salinities in the estuary. Oysters depend on naturally lower salinities and oyster beds decline with lower flows. The same is true of the commercial shrimp and crab fisheries in the Bay. Lower flows mean higher salinity, which harms oyster, shrimp, and fish production — direct adverse impact on the local economy. Atlanta's recent drought problem was temporarily ameliorated by heavy rains in the winter of 2008–2009 but drought will recur. More severe droughts than the most recent one can be expected in the future because there have been many droughts in the southeast as intense as the recent one, but lasting for a decade or longer.[21]

River ecologists have demonstrated that the pattern of natural flow is a "master variable" that helps create the extraordinary productiveness of rivers.[22] The effect of urban and agricultural withdrawals of water must be considered together with the consequences of damming, diking, and channelization of a river system. Species diversity and biological productivity are often harmed by reducing the variability in flows that mark natural rivers. Reservoirs upstream of the

Apalachicola have harmed the natural flow regime and have effects that trace their way through the physical, chemical, and biological foundation of the flowing systems.

Flow regulation practiced by the Army Corps of Engineers at several large reservoirs is one factor in the equation of loss. Others are urban, agricultural, and industrial withdrawals, changes in rainfall patterns, and evaporation from more than 25,000 artificial ponds and reservoirs in the basin.[23] Four-fifths of the small impoundments in the basin are in Georgia. Most river and floodplain species reproduce primarily in the spring and summer, and many species are especially vulnerable during dry conditions. The unfortunate effect of upstream water management is that reduced flows from upriver happen just when the most critical biological processes in the system occur.

The U.S. Geological Survey confirms that water-level declines over the last 50 years have damaged the Apalachicola River system.[24] The harm has been done by a combination of reduced flow from upstream activities and channel erosion of the bed and banks of the Apalachicola River. The large federal reservoirs upstream catch the sediment that otherwise would move downstream. That missing sediment is vital to maintaining the health of the Apalachicola River. Without the natural sediment, the bottom of the Apalachicola River has become scoured out and artificially lowered. Navigation projects of the Corps of Engineers also lowered the river bottom by decades of maintenance dredging and by modifying the natural river channel. Because the bottom of the river channel is lower and the channel is wider, the level of the river surface has dropped. That means that even if flows were returned to natural conditions, water levels would still be lower than natural levels. The greatest water level decline is at low and medium flows. In the upper Apalachicola River, the water level in spring and summer of dry years is several feet lower than it was prior to the 1950s. The damage to the floodplain forest caused by these water-level declines is extensive. An estimated 4.3 million trees have been lost, with swamp species such as tupelo and cypress being the hardest hit.[25] As a consequence, the floodplain of the river, which is its biological engine, is being starved of water.

The potentially damaging effects of declining water levels have been known for decades. As far back as 1983, Governors Bob Graham of Florida, Joe Frank Harris of Georgia, and George Wallace of Alabama signed an agreement to work together to solve basin problems. Nonetheless, the Army Corps of Engineers changed their reservoir policies in 1986 to provide additional water supply services for Atlanta. This action reduced flows downstream for other water users during drought, including those in Alabama and Florida. Both Alabama and Florida objected to this new allocation of reservoir water for urban uses at the top of the basin. In 1990, Alabama filed a federal lawsuit and Florida intervened on behalf of Alabama's case. The court case was not resolved then because the states agreed to put it aside and cooperate in a multimillion-dollar investigation of the basin. No binding allocation agreement was reached after years of study. Instead, the three states agreed to discuss the disagreement in another forum, this time in an interstate compact. At the states' request, Congress enacted the Apalachicola-Chattahoochee-Flint (ACF) River Compact on November 20, 1997. Unlike most compacts, the arrangement had no allocation of flows. The three states and the federal government agreed only to try to arrange for a mutually acceptable water allocation. The unusual compact provided that it would self-destruct unless the four parties agreed to an allocation by December 31, 1998.

Two people served as nonvoting federal commissioners during the life of the compact, along with appointees from the three states. From Florida's perspective, neither of the federal commissioners serving as chair of the negotiations could plausibly claim neutrality. President Bill Clinton appointed Lindsay Thomas, a former Congressional representative from Georgia who served later as the president and CEO of the Georgia Chamber of Commerce. Later, President George W. Bush appointed Alec Pointevent, who was the general chair of Georgia Governor Perdue's gubernatorial campaign and co-chaired his inauguration committee.

The deadline for unanimous agreement on an allocation came and passed without a settlement. The compact included no explicit provision for extending the deadline for reaching an agreement, but the four parties nonetheless unanimously agreed to an extension 13 separate times. In the end, Georgia would not agree to "any specific minimum flow conditions on the river system."[26] Feelings ran hot during the final stages of negotiation, just before the ACF compact lapsed. The secretary of the Florida Department of Environmental Protection looked to Winston Churchill and World War II for an analogy: "We shall never surrender either, because the stakes are high."[27] Florida did not surrender but neither did it prevail in compact negotiations. The compact, after the many extensions, was allowed to expire in 2004.[28] The downstream states of Florida and Alabama reactivated their lawsuits.

The Apalachicola remains the most critical issue for the Northwest Florida Water Management District and the Florida Department of Environmental Protection. However, the region is growing fast and soon will face the same panoply of daunting water management issues confronting the peninsular districts of Florida. "Getting the water right" in northwest Florida will require action from the district and state government in three key areas: fostering a culture of water sustainability, raising the millage cap, and winning the political battles on the ACF system.

FOSTERING A REGIONAL CULTURE OF WATER SUSTAINABILITY

Sustainability must be sought in order to be found. Unfortunately, the political opposition that deprived the Northwest Florida Water Management District of a necessary tax base in 1976 was displayed again in 1997 when the legislature upgraded the state's wetland protection statute – except in the Northwest Florida Water Management District. As a local legislator expressed it in 1996, "We've got plenty of wetlands, but we don't have very many Wal-Marts."[29] North Florida legislators exempted the district from the wetland regulation improvements for a

Figure 6.1 *It is necessary for the water management districts to establish minimum flows and levels to protect species like the Florida manatee*

decade. It wasn't until 2010 that the region had the same level of wetlands protection as the rest of Florida.

The Northwest Florida Water Management District is assigned by state law to share in an interstate water management responsibility. George Willson, a member of the Northwest Florida Water Management District governing board from 1991–1999 and an observer of their board meetings for 20 years, believes that the district has to act less "like a big rural county commission" and more like a full-featured water management district willing to conduct effective regulation.[30] The district can help build a culture of sustainability in the many watersheds of the region that feed into the northern Gulf of Mexico. It is disappointing, therefore, that they are the only water management district in the state that has not adopted a single minimum flow and level for any water body. (The district did adopt a "reservation" of water for the Apalachicola River as a litigation strategy in the interstate lawsuits, but that has little effect on Florida water users.)

RAISING THE MILLAGE CAP

Many statewide commissions over the last three decades have recommended that the Northwest Florida Water Management District millage cap be raised to the level of the other water management districts. However, the legislature has always refused to put it on the ballot for voters to decide. Historically, the transfer of state revenues from other parts of Florida to North Florida was a prime objective of the legislative Pork Choppers described in Chapter 2. Today's tax arrangements for the Northwest Florida Water Management District continue that unsustainable heritage. As a revenue supplicant before the legislature, the Northwest District cannot take stands that upset development interests because the stream of state dollars might be interrupted if it did.

Only the governor of Florida, prompted by the people of northwest Florida, could make a difference in the millage cap. To date, no governor has made that a priority because the residents have not made it one. If sustainability is a goal, then adequate taxation to achieve that goal will have to occur.

WINNING THE LEGAL AND POLITICAL BATTLES ON THE ACF SYSTEM

The state of Florida spent two decades in studies of the ACF basin and a failed effort at an interstate compact. Although the science of the rivers and bay was improved over that period, it might have been better to simply allow the lawsuits to proceed.[31] There seems little reason to conduct serious interstate negotiations unless legal incentives change favorably. Sustainable interstate water management for the Apalachicola may be possible only if Florida scores a very clear win in the lawsuits on the ACF and is solidly backed up by the federal government. That will require federal judges to decide that the Apalachicola is worth saving; a succession of Florida governors will have to make the river and bay a priority; and the president of the United States will have to direct that federal agencies align with those same goals.

Florida and Alabama won a major legal victory in 2009, which may be the breakthrough needed for the states to take new approaches to solve this long-standing dispute.[32] The complicated tri-state litigation was assigned by the federal court system to U.S. District Judge Paul Magnuson from Minnesota. He previously had served in the federal court system as chair of the Committee on International Judicial Relations, chair of the Advisory Council of the International Association for Court Administration, and chair of the Committee on Administration of the Bankruptcy System. He also oversaw Endangered Species Act litigation involving the Corps' action on the Missouri River, where he came down on their side of the issue.

At oral arguments in Jacksonville about the case, the judge gave warning that he expected to render a clear decision, for one side or the other: "Somewhere, some place, something's got to come to the attention of people. We're dealing with potential major environmental issues. We're dealing with potential turn-the-lights-off issues. We're dealing with potential turn-the-water-off issues."[33] He also was impatient with the statement by a Corps attorney that a new water management plan for Lake Lanier would not be completed until 2012: "What in blazes are you doing? It's a situation that simply cannot be permitted to continue."

This foreshadowed Judge Magnuson's lengthy ruling on July 17, 2009 about whether Atlanta had a legal right to use Lake Lanier for water supply. According to the federal Water Supply Act, a reservoir built for other purposes could not undergo major structural or operational changes for water supply purposes without approval of Congress. Magnuson reached what he termed the "inescapable conclusion that water supply, at least in the form of withdrawals from Lake Lanier, is not an authorized purpose of the Buford project."[34] He ruled that the failure of the Corps to seek Congressional authorization was an abuse of their authority and that they were in violation of the Water Supply Act. This mattered because, as he noted, "Even if annually the average flows are reduced by only a small amount, as the Georgia parties argue, the actual variation in flows can wreak havoc on the downstream uses of the water."

The judge gave the Corps and local interests three years to seek Congressional authorization for water supply operations. After that, in what he called a "draconian" result, Atlanta must take no more water out of the system than it did in the mid-1970s. He urged a more far-sighted water management approach:

> Too often, state, local, and even national government actors do not consider the long-term consequences of their decisions. Local governments allow unchecked growth because it increases tax revenue, but these same governments do not sufficiently plan for the resources such unchecked growth will require. Nor do individual citizens consider frequently enough their consumption of our scarce resources, absent a crisis situation such as that experienced in the ACF basin in the last few years. The problems faced in the ACF basin will continue to be repeated throughout this country, as the population grows and more undeveloped land is developed. Only by cooperating, planning, and conserving can we avoid the situations that gave rise to this litigation.[35]

This ruling was only on the issues relating to the lack of legal authority to use Lake Lanier for public water supply in the northern end of the interstate watershed. Judge Magnuson later decided that environmental issues, like those arising from the Endangered Species Act, were important but that he would not set aside the current endangered species findings of the U.S. Fish and Wildlife Service in the southern end. Florida may appeal that second ruling, which discounted concerns about species endangerment.

Judge Magnuson is correct that the states of Georgia, Florida, and Alabama must collaborate and find a way to optimize the resources of the river system. Long-term sustainability has to be the overarching goal. There must be quantitative limits for how much water upstream states can remove from the living Apalachicola River (and not merely a negotiated agreement on estimated future flows in the river itself). Legal victories are essential in reaching this goal but will mean little if the southeastern states do not agree on managing their common resources for future generations. However, in the end, the three states must find ways to manage cooperatively the shared water resources. Ironically, a court order may be the means by which cooperation among the states is achieved.

NOTES

1. Oral History Program. (2006) Don Duden, 2 November. It should be noted that Doug Stowell, an aide to Governor Reubin Askew at the time and later the general counsel for the Northwest Florida Water Management District, believed that the legislator most responsible for the low millage rate for NWFWMD was state representative Billy Joe Rish from the city of Port St. Joe (Oral History Program, November 18, 2005).

2. Blaustein, Richard J. (2009) "Biodiversity hotspot: The Florida panhandle," *Bioscience*, vol. 58, no. 9, pp784–790

3. Thousand Friends of Florida. (2010) "The Florida Panhandle Initiative: Protecting biodiversity in northwest Florida," www.1000friendsofflorida.org/Panhandle/Biodiversity.asp, accessed 3 July 2010

4. Northwest Florida Water Management District. (2008) *Emerging Waters: Springs of Northwest Florida*

5. Northwest Florida Water Management District. (2009) *Northwest News & Updates*, vol. 1, no. 3, p1

6. Manning, Phillip. (1997) *Orange Blossom Trails: Walks in the Natural Areas of Florida*, John F. Blair Publishing, Winston-Salem, NC, p196

7. "St. Joe Company," http://ir.joe.com/overview.cfm, accessed 10 July 2010

8. Ziewitz, Kathryn, and June Wiaz. (2004) *Green Empire: The St. Joe Company and the Remaking of Florida's Panhandle*, University Press of Florida, Gainesville, FL, p299

9. Pitman, Craig "Billy Buzzett, Scott's Choice to Lead Growth Management, Has Long Family History in Florida" *St. Petersburg Times*, January 8, 2011.

10. Northwest Florida Water Management District. (2008) *2008 Water Supply Assessment Update*, p3–23

11. Northwest Florida Water Management District. (2009) "Planning helps expand district water supply," *Northwest News and Updates*, vol. 1, no. 1, p1

12. Northwest Florida Water Management District. (2008) *Land Acquisition Work Plan*

13. Pacific Institute. (2006) *A Review of Water Conservation Planning for the Atlanta, Georgia Region*, Pacific Institute, Oakland, CA

14. Mary Ann Dickinson, president and CEO, Alliance for Water Efficiency, letter to Alice Miller Keyes, Georgia Department of Natural Resources, January 31, 2009

15. Campbell, Sarah Fay. 2010 "Smith's water bill awaits Perdue's approval," *Times-Herald* (Newnan, Georgia), 21 March

16. Dellapenna, Joseph W. (2005) "Georgia water law: How to go forward now?" *Proceedings of the 2005 Georgia Water Resources Conference*, University of Georgia, Athens, GA, April 25–27, 2005

17. Feldman, David Lewis. (2007) *Water Policy for Sustainable Development*, The Johns Hopkins University Press, Baltimore, MD, p138

18. Golladay, S. W., D. W. Hicks, and T. K. Muenz. (2007) "Stream flow changes associated with water use and climatic variations in the lower Flint River basin, southwest Georgia," *Proceedings of the 2007 Georgia Water Resources Conference*, University of Georgia, Athens, GA, March 27–29, 2007

19. Georgia Department of Natural Resources, Environmental Protection Division. (2009) *Georgia's Water Conservation Implementation Plan*, p38

20. Georgia Water Coalition. (2008) *2008 Annual Report*, p8

21. Seager, Richard, Tzanova, Alexandrina, and Nakamura, Jennifer. (2009) "Drought in the southeastern United States: Causes, variability over the last millennium, and the potential for future hydroclimate change," *Journal of Climate*, vol. 22, pp5021–5045

22. Richter, B. D., R. Mathews, D. L. Harrison, and R. Wigington. (2003) "Ecologically sustainable water management: Managing river flows for ecological integrity," *Ecological Applications*, vol. 13, no. 1, pp206–224

23. Ignatius, Amber R. (2009) "Big water, little water: Identification of small and medium-sized reservoirs in the Apalachicola-Chattahoochee-Flint River basin with a discussion of their ecological and hydrological impacts," MS thesis, Department of Geography, Florida State University, Gainesville, FL

24. Darst, Melanie R., and Helen M. Light. (2008) *Drier Forest Composition Associated with Hydrologic Change in the Apalachicola River Floodplain, Florida.* Prepared in cooperation with the Florida Department of Environmental Protection and the Northwest Florida Water Management District, Scientific Investigations Report 2008-5062

25. Livingston, Robert J. (2008) *Importance of River Flow to the Apalachicola River-Bay System,* Report to the Florida Department of Environmental Protection. Department of Biological Science, Florida State University, Gainesville, FL

26. Sherek, George William. (2005) "The management of interstate water conflicts in the twenty-first century: Is it time to call uncle?" *New York University Environmental Law Journal,* vol. 12, pp764–827

27. Struhs, David B. [Secretary, Florida Department of Environmental Protection]. (2002) *Statement on Exiting the Apalachicola-Chattahoochee-Flint Rivers Water Allocation Compact,* Florida Department of Environmental Protection, Tallahassee, FL

28. Leitman, Steven.(2005) "Apalachicola-Chattahoochee-Flint basin: Tri-state negotiations of a water allocation formula," in John T. Scholz and Bruce Stiftel *Adaptive Governance and Water Conflict,* Resources for the Future, Washington, DC, p88

29. Cushman, John H. (1996) "Wetlands are turning to asphalt, filled in a parcel at a time," *New York Times,* 27 October

30. Oral History Program. (2007) Interview of George Willson, 27 August

31. Ruhl, J. B. (2005) "Water wars, eastern style: Divvying up the Apalachicola-Chattahoochee-Flint river basin," *Journal of Contemporary Water Research and Education,* issue 131, pp47–54

32. Jordan, Jeffrey L., and Aaron T. Wolf. (2006) *Interstate Water Allocation in Alabama, Florida, and Georgia,* University Press of Florida, Gainesville, FL

33. Rankin, Bill. (2009) "Army Corps criticized by Lake Lanier judge," *The Atlanta Journal-Constitution,* 11 May

34. U.S. District Court, Middle District of Florida. (2009) Tri-State Water Rights Litigation, Case No. 3:07-md-01 (PAM/JRK), Judge Paul Magnuson, Memorandum and Order, July 17, 2009

35. Ibid.

CHAPTER SEVEN

Southwest Florida:
A Truce in the "Water Wars"

"**S**wiftmud" (the Southwest Florida Water Management District, SWFWMD) was created a little more than a decade after the Central and Southern Florida Flood Control District was in 1948, and provides another demonstration of the interplay of Florida climate and politics. Perhaps even more importantly, the history of the Southwest Florida Water Management District demonstrates how the Florida water management system is capable of major institutional transformation.

The origins of the district are in the caprices of Florida's climate. The 1956–1957 period was one of the driest in the state's history but was followed immediately by wet years in the region in 1958, 1959, and 1960. In March of 1960, a stationary front dumped an average of more than 10 inches of rain over 10,000 square miles of central Florida, flooding more than 1 million acres. Some areas received up to 27 inches of rain during the March 15–18 downpour.[1] On April 13, 1960, John W. Wakefield, the director of the Florida Department of Water Resources, testified to a United States Senate appropriations committee about flooding in west-central Florida. The state wanted Congress to direct the U.S. Army Corps of Engineers to investigate the region's water problems, including the devastating floods just a month before the hearing. The Corps study would "bring encouragement and hope to a thoroughly disheartened and confused segment of an otherwise progressive and rapidly growing state."[2]

The plea for help was better timed than Director Wakefield could know. In July, 1960, 20.6 inches of rain fell in Tampa, a record still not exceeded. In response to regional flooding, local governments organized a committee chaired by Brooksville banker Alfred ("Fred") McKethan.[3] McKethan was once in the running for governor, was the former chair of the State Road Board, and was later president of the Florida Bankers Association, and he worked closely with state officials on this issue. On September 2, Wakefield sent recommendations to the

Central and Southwest Florida Water Management District Steering Committee on proposed legislation for a new regional agency. His memo recommended:

> That the name of the district be the 'Southwest Florida Water Control (or Water Management) District.' The word 'Central' has been omitted to prevent confusion between the Central and Southern Florida Flood Control District, and the new organization. The words 'Flood Control' have been omitted to point up the basic philosophy of comprehensive water management for *all* legitimate purposes.[4] (emphasis in original)

This memorandum was timed even better than his Congressional testimony in the spring. Only 8 days after Wakefield sent the memo, Hurricane Donna hit the Florida Keys and then smashed into southwest Florida on September 11, 1960, causing an 11-foot storm surge. Donna then moved northeastward, crossing the middle of the peninsula before entering the Atlantic Ocean to the south of Jacksonville. It made landfall again in North Carolina and did significant damage as far north as Long Island, New York. Hurricane Donna did so much damage that its name was permanently retired from future use. For west-central Florida, Donna brought even worse flooding to a region already primed for action. In Hernando County near Brooksville, for example, one flooded area was a mile wide and more than 5 miles long. Across the region, 700,000 acres were inundated, causing $11 million in damages. The Hillsborough River, which runs through Tampa, flooded more than a hundred homes. The Florida legislature immediately created the Southwest Florida Water Management District in the spring of 1961, as recommended by the steering committee and the state Board of Conservation. The magnitude of flooding was probably not greater than in previous floods, such as one in 1933, but intense development since earlier large floods magnified the property damage.

Governor Farris Bryant named Fred McKethan, a political ally, to chair the new water management district. John Wakefield was assigned from the state to serve as acting executive director for a short while to help get the district up and running. The first meeting of the governing board was held on August 28, 1961, in the Cabinet room in Tallahassee, which made it convenient for the governor to advise the new district on priorities. Bryant agreed that "water management rather than flood control is the key to a successful long range program."[5] The governor also commended the new district for "maximum decentralization," a reference to the unique system of district "basin boards" in the authorizing legislation. He said it was "vital that the tax monies be spent in those localities where the public is willing to participate in terms of benefits received." (As shown later in this chapter, that advice has been followed throughout the history of the district.)

One of McKethan's first decisions as chair was to arrange for the headquarters of the new water management agency to be only a few blocks from his office in rural Brooksville, even though it was many miles away from the geographic or population center of the new district. All of the board members were his friends and allies. When the payroll account for the district was established, it was deposited in McKethan's bank in Brooksville.[6] The headquarters are still in Brooksville (although there are now branch offices in several other cities).

The Southwest Florida Water Management District would, in the next decades, undergo major transformation as the region grew to be the home of about a quarter of the state's population and 100 cities and county governments. The fiscal and staffing resources brought to bear today by this district are much larger than those available to either the Suwannee River or Northwest Florida Water Management Districts. This district's annual budget is approximately $300 to $400 million a year and it employs almost 800 people. It has the resources, as McKethan said, to "do big jobs and do them well."[7]

Although it was created principally as the local sponsor of an Army Corps of Engineers flood control project, the Southwest Florida Water Management District would soon become enmeshed in major water supply battles. Rather than focusing on building water supply projects, it would become a state and national leader in promoting water use efficiency. The Swiftmud experience illustrates – better than the lightly populated Suwannee River or Northwest Florida districts – how a water management agency confronts issues in a region with dense urban settlements, intensive agricultural irrigation, and major industries. Today, the state's longest-running "water wars" in the Tampa Bay area have reached an armistice, in part due to Swiftmud's ability to evolve over time.

THE WATERS OF THE DISTRICT

The water resources of the Southwest Florida Water Management District reach across a 10,000-square-mile region. The western side of the peninsula faces the Gulf of Mexico, and 13 major rivers of the district drain much of it. Close to the northern boundary of the district, just south of the Suwannee River, some of the largest springs in the world form a span of unique coastal rivers. The Weeki Wachee, Crystal, Homosassa, and Chassahowitzka Rivers are all relatively short streams comprised mostly of discharges from large springs. All of them make rapid transitions from pure freshwater at their headsprings to saltmarsh estuaries in a distance of only a few miles. For example, the Weeki Wachee runs only 8 miles to the Gulf of Mexico from its first-magnitude headspring. The main spring vent is 45 feet deep. Average spring flow in recent years has been about 150 to 200 cubic feet per second, derived from a springshed of more than 250 square miles. Groundwater withdrawals have significantly reduced spring flow.[8]

The rivers of the Springs Coast face challenges stemming from population growth. Fertilizers applied to home landscapes and golf courses have dramatically increased the concentrations of nitrates in the springs. Taken as a whole, the 900 cfs of water discharging from springs in the region carry more than 360 tons of nitrates to their estuaries each year. If nitrate concentrations are not reduced, the rivers and estuaries will see increasing algal blooms and reduced ecosystem diversity.[9]

Farther south, in the geographic and population center of the district, is a peninsula on the peninsular state: Pinellas County, on the west side of Tampa Bay. The hydrologic isolation of Pinellas County has no doubt contributed to the long-lasting water wars in west-central Florida. In the early twentieth century, the rapidly growing population of St. Petersburg in southern Pinellas County quickly exhausted the locally available supplies of water. In 1940, St. Petersburg purchased Weeki Wachee Springs, 55 miles up the road, as a water supply source for the distant future. A private company opened an underwater "mermaid show" at Weeki Wachee in 1947, which continues today. St. Petersburg retained ownership of Weeki Wachee until 2001 but never did build a pipeline that far to the springs. The Southwest Florida Water Management District purchased the struggling old-style Florida attraction in 2001. Today, Weeki Wachee Springs is managed as a state park – the only park in the world that employs mermaids.

Eighteen percent of the Southwest District is wetlands, despite a century of drainage projects. The district includes the Green Swamp area in its northeastern region, which is the state's second largest wetlands system (after the Everglades) and the headwaters of four large Florida rivers: the Withlacoochee, Hillsborough, Ocklawaha, and Peace. The 870 square miles of the Swamp are not all wetlands; in fact, it is a complex mosaic of marshes, forested wetlands, and uplands. Part of the Swamp is known for high rates of groundwater recharge, which feeds the Floridan Aquifer of much of the region.

In the 1960s the district intended to build very large impoundments in the Swamp to hold back floodwaters and prevent flood damages downstream. That approach was in accord with the emphasis on structures that dominated water management policies in that era. Fortunately, the authorized flood control project was reconsidered and eventually shelved before it was constructed. The district was also instrumental in the designation of the Green Swamp as a state Area of Critical State Concern, and it owns more than 150 square miles of it. Today, the Green Swamp still stores floodwaters, which trickle slowly down the rivers but not from artificial impoundments.

On the central Gulf coast of the district is Tampa Bay and the metropolitan region that surrounds it. Human activities have eliminated some of the original natural features of the bay. About half of the natural shoreline of Tampa Bay has been lost to development. Nonetheless, very valuable resources remain, and improvement of the estuary is a priority for the Tampa Bay National Estuary Program. There have been some notable successes. For example, water quality is better today than it was 20 years ago and Tampa Bay has several thousand more acres of seagrass than it did in 1982.[10]

At the southern end of the district, rivers are not primarily spring-fed but are the more common type of Florida blackwater streams, stained a natural tea color from vegetation. The Myakka River, for example, runs 66 miles from its headwaters to Charlotte Harbor. Wetlands comprise 55 percent of the 600-square-mile Myakka watershed. The legislature designated it a state Wild and Scenic River in 1985 because of its "outstandingly remarkable ecological, fish and wildlife, and recreational values which are unique in the State of Florida." (Section 258.501, F.S.)

However, the state law making the Myakka a Wild and Scenic River did not protect it against all threats, including excessive agricultural irrigation in the watershed. The water management district created a problem for the Myakka River by its own permitting of agricultural irrigation in the watershed. So much water is used for irrigating vegetable farms in the Flatford Swamp basin of the

Myakka River that runoff continually seeps into streams and the water table downstream. That has made the Swamp much wetter during the dry season than it is naturally.[11] As a result, much of the hardwood forest in Flatford Swamp, which is owned by the district, has died due to high water conditions. Even Myakka River State Park downstream has suffered the loss of trees.

Eight rivers are used in the Southwest District for public or industrial water supply, which is more than in any other water management district. The Hillsborough River, Braden River, Manatee River, Shell Creek, and Myakkahatchee Creek all have impounding structures in their river channels. The Peace, Alafia, and Little Manatee Rivers have not been impounded, but substantial quantities of water are removed from them and are stored in off-line reservoirs built with levees aboveground. Most of the dams on rivers in the district date back to the 1950s and 1960s and probably could not meet environmental standards if they were proposed today.

The district's rivers are the lifeblood of its estuaries, which are hatcheries for many ecologically and commercially important species. Sarasota Bay, south of Tampa, is also a National Estuary, as is Charlotte Harbor at the southern edge of the district. The Charlotte Harbor estuary is the largest in the district; its 270 square miles make it the state's second largest open-water estuary.

Almost a quarter of the state's lakes are located in the Southwest District. Some of them were severely abused in the heyday of unrestricted Florida development in the early and mid twentieth century. Lake Hancock, at the headwaters of the Peace River, was damaged by phosphate mining, agricultural runoff, and municipal pollution. Polk County's director of natural resources said, "Just about anything bad you can do to a lake, we've done to this lake over the last 150 years."[12] However, the district is embarking on a $200 million project to clean up the 4,500-acre lake and raise its water level by a foot. This would both enhance the lake and make water available in the dry season to feed the headwaters of the Peace River.

In the late nineteenth century, the Peace River basin section of the district became an agricultural center of cattle ranching and orange groves. The region's history changed dramatically when marketable concentrations of phosphate ore were discovered there in the 1880s. By 1900 more than 200 companies were mining phosphate in central Florida.[13] The state song in the early twentieth century paid tribute to the importance of phosphate mining as part of the state's economy: "Thy golden fruit the world outshines, Florida, my Florida. / Thy gardens and thy phosphate mines, Florida, my Florida." Phosphate mining in Florida is conducted mostly in the Southwest Florida Water Management District. (There is also one mine in the Suwannee River Water Management District.) Today, about 65 to 75 percent of U.S. mining and 25 percent of world mining of phosphate rock occurs in Florida.[14] In effect, Florida is being mined to support agriculture in other states and countries. Excessive amounts of phosphate fertilizer applied to American farms, and inadequate control of farm runoff, help create biological dead zones like the 8,000-square-mile dead zone at the mouth of the Mississippi River in Louisiana. Worldwide, rivers now carry more than twice as much phosphorus to the ocean compared with pre-mining times.[15]

Phosphate mining is often preceded by wetland destruction, and water use in this activity is immense. This part of the Floridan Aquifer used to be called the Big Red Hole on maps of groundwater levels in the region. Phosphate mining and agriculture used so much water in the Bone Valley region of the Southwest District that it had dried up Kissengen Spring in Polk County by 1950. In recent decades, miners dramatically reduced the amount of water they used. From 1970 to 1995, the amount of water used per ton of phosphate ore extracted declined from 3,500 gallons to less than 550 gallons. Most mines today use even less.[16] The district pressed the companies to conserve water but the principal reason why the companies used less water was that water conservation saved them money. The reduction in water use, plus the mining-out of some phosphate resources, has helped improve groundwater levels in the Floridan Aquifer east of Tampa. Management of these massive mining operations, as well as the mined-over areas, will be a concern of the district for decades to come.

AQUIFER PROBLEMS

The Floridan Aquifer is the principal source of groundwater in the Southwest Florida Water Management District. Water users in the district withdraw more groundwater per square mile than users in any other water management district in Florida or in any other state as a whole. Overpumping of aquifers is one of the district's greatest concerns because it can draw in saltwater from the coast. The district designated a portion of the Floridan Aquifer southeast of Tampa as the Southern Water Use Caution Area (SWUCA) due to concerns about too much pumping. Surface features such as lakes and wetlands were being drained in the eastern part of the SWUCA; saltwater to the west was being drawn in near the Gulf coast. It took 15 years of planning, rule writing, and strategy development before the district could finally put in place a groundwater recovery strategy on January 1, 2007.

The legal battle with local governments and water users about limiting groundwater withdrawals produced the longest administrative hearing in Florida history.[17] Swiftmud won on most of the legal points, including the ability to set minimum aquifer levels to reduce saltwater intrusion. The precedent set by limiting withdrawals from a regional water resource was important for other water management districts. The SWUCA influenced the setting of groundwater withdrawal caps in the Orlando region in 2006, which were adopted jointly by the Southwest Florida Water Management District, the St. Johns River Water Management District, and the South Florida Water Management District.

By 2025, if the SWUCA strategy is successful, inland lakes in the Highlands Ridge area of Florida will have met their established minimum levels and the Peace River will be flowing at a pace that also meets the minimum flows adopted by the district. Achieving this goal will be very challenging, in light of the hundreds of millions of gallons used by permittees today and potential additional future demands. At best, however, saltwater intrusion will be reduced to a pace of a few hundred feet inland per year. Eliminating saltwater intrusion would require pumping reductions so large that the Southwest District has already deemed them infeasible. Eventually, wells by the coast that become salty will have to be moved inland, be drilled deeper, or be abandoned.

BASIN BOARDS

The Central and Southwest Florida Water Management District Steering Committee that proposed the creation of the Southwest Florida Water Management District was made up of local government representatives. They wanted to maintain a direct say in the district's actions within their own jurisdictions.[18] Consistent with most of the political history of Florida, voters and politicians had little trust in state institutions. For the Southwest District, that produced a more decentralized and cumbersome structure than that of any other water management district. There are seven separate basin boards under the main districtwide governing board, each having at least three members appointed by the governor and confirmed by the senate. In addition to the Southwest District, only the South Florida Water Management District has any basin boards – only one, covering the Big Cypress basin on the southwest side of that district. The Southwest District must carry the administrative and political overhead of dealing with its seven basin boards and more than three dozen basin board members, in addition to the 13 members of the main district governing board. Every year the district has monthly governing board meetings plus several dozen additional basin board meetings and basin board workshops.

The basin boards are most interested in water management activities and projects within their own boundaries, and have independent authority to levy property taxes to implement their priorities. In the early days of the district, the basin boards concentrated on funding structural facilities like those in the Four River Basins Project (described in the next section). By the 1990s their focus had shifted to broader purposes, including floodplain management, water conservation, and the development of alternative water supplies. Taken together, the basin boards and the main governing board have the ability to raise many millions of dollars each year from property taxes.

The Southwest District's governing board has 13 members – 4 more than any other water management district – to oversee the district. Having the largest governing board again reflects the desire of local interests in the Southwest District to have a voice in how the district approaches their water management goals. This intense localism has at times threatened the district's stability. In the 1980s, the district's main governing board considered reducing the number of basin boards to only three. Tampa-area legislators also made an effort to move the Southwest District headquarters from Brooksville to Tampa by holding up state appropriations to the district.[19] Disputes between the basin boards and the main governing board forced Governor Bob Graham to intervene personally as a peacemaker in 1984. He threatened that any board member not committed to settling the dispute "needs to reconsider the prospect of continuing to serve."[20] No such governance crisis has occurred since that eruption.

THE FOUR RIVER BASINS PROJECT

Flood damage arising from Hurricane Donna and other recent rainfall events inspired the project but the Corps and the district also looked more broadly to water storage, saltwater intrusion, fish and wildlife habitat, and recreation benefits. The Corps project included the Green Swamp; the Hillsborough River (draining about 690 square miles), the Ocklawaha River (draining about 2,100 square miles to the St. Johns River), the Peace River (with its 1,900-square-mile watershed), and the Withlacoochee River (draining about 1,980 square miles). In the spring of 1960, the chief of the Army Corps of Engineers predicted that a detailed study of the watershed's problems could not even begin until 1962 and would have to be completed before Congress could consider action.[21] He underestimated the pressure that would be applied by Florida politicians. On October 23, 1962, only 14 months after the first meeting of the new Southwest District Governing Board, Congress authorized the Army Corps of Engineers to construct the Four River Basins Project, with the Southwest District as the local project sponsor. The Four Rivers Basin Project was a nearly irresistible financial deal for the district; Swiftmud was expected to contribute land but pay only 14 percent of construction costs.[22]

Although the new Southwest Florida Water Management District incorporated some features of the Central and Southern Florida Flood Control District, the district's executive director regarded the flood control district in south Florida as doing things the "old way," whereas the new district wanted to pioneer new approaches. Staff of the Southwest Florida Water Management District regarded themselves as the virtuous "white hats" of water management while the Central and Southern Florida Flood Control District represented to them the "black hats."[23] According to Dale Twachtmann, the district's early executive director,

> John Wakefield had explained to me that the reason that the name should be Water Management District was because we should not be talking just about flood control. The one in south Florida was called Central and Southern Florida Flood Control District. He didn't think that should be the case. This district should be formed in such a way that you could manage water whether there be drought or flood. That was a far more sensible thing. I thought, well, that's terrific, makes good sense to me. So SWFWMD was the first one named Water Management District.[24]

The transformation of the Corps project by the new district is a leading example of such new approaches. The original Four River Basins Project contemplated 12 separate reservoirs, including 2 very large "conservation pools" created by 30 miles of constructed levees. The district promptly used its new taxation powers to acquire large tracts of land intended for the impoundments. As the time drew nearer for building the associated levee structures, however, doubts arose about whether they were desirable in light of new understandings of the natural values of the Green Swamp. Swiftmud concluded in the early 1970s that permanent pools of water stored behind the levees would cause unacceptable environmental damage. District staff recommended lower levees that would still impound water in times of high rainfall, but for shorter periods. The district staff feared that if the flood control project were entirely dropped, the Green Swamp would be developed because the district might lose its justification for owning land in the Swamp. Their environmental report said, "Without implementation of the project, the natural environment of what remains of the Green Swamp will be lost and the people of Central Florida will have the dilemma of having to choose between periodic flooding or channelization of their native streams."[25] Nonetheless, the district proceeded to redesign the Four River Basins Project without so many structures and in the end managed to retain the land, too.

More than a decade was spent in redesigning the original project, which had been heavily dependent on structural facilities. The district considered lower levees, shorter levees, and levees in somewhat different locations to reduce environmental damage while still providing flood control benefits. By the late 1980s, the district had abandoned the original impoundment idea. For example, Corps and the district originally intended to build a Green Swamp Flood Detention Area but the 110,000-acre area is now managed by the district as the Green Swamp Wilderness Preserve.

The Green Swamp and other land acquisition projects comprise a large fraction of the 430,000 acres owned by the district today. Even without levees, the natural Green Swamp and other low-lying areas downstream provide water storage and flood attenuation benefits. However, not all of the Four River Basins Project was transformed from the original conceptions. The district currently maintains 81 water conservation structures, salinity barriers, and flood control structures, and more than 50 miles of canals and levees. Many of these structures were inherited from projects built before 1960.

The district's largest flood control facility today is the Tampa Bypass Canal. In cooperation with the Army Corps of Engineers, the district reduced flooding in downtown Tampa by building a bypass canal to receive floodwaters rushing down the Hillsborough River in wet times. They found a convenient way to build that system by digging an overflow, or bypass, canal from the Hillsborough River to a parallel small stream. The district and the Corps destroyed that small river as thoroughly as any in Florida. Six Mile Creek, and its downstream reach called Palm River, were "completely excised" in the early 1970s to create the floodway.[26]

Although the district channelized Six Mile Creek early in its existence, it was deterred from destroying part of Tampa Bay. Making part of Tampa Bay a freshwater lake was an old dream in the region. In 1951, the Army Corps of Engineers proposed that the upper northwest section of Tampa Bay, called Old Tampa Bay, be converted into the Upper Tampa Bay Fresh Water Lake.[27] Dale Twachtmann, the district's executive director, supported another impoundment project on Tampa Bay, in part because the freshwater in the lake would tend to prevent saltwater intrusion into wellfields. He and others at the district were stunned by the level of opposition generated by a nascent environmental movement:

Biologists all of a sudden started hearing about it, and Dr. William Taft from the University of South Florida came out, and he was very bold and very loud and he started talking about this water management district and this dumb idea, and it's going to be a terrible thing. Then he got people from South Florida and from the Game and Fish Commission and others, and pretty soon we were having these meetings and these people were showing up and saying, you don't know what you're doing. A fresh water environment taking over a salt water environment means years of smell and stink, and it's too shallow and it's too warm. You're going to destroy the nesting grounds for the brown pelican, and you're going to bring in different kinds of grasses and the old grasses won't grow, they'll die. There are certain fishes that are going to die. All the fresh water species are going to die, and the salt water species are going to be hard to get there because you've got this all blocked off. It just went on for months. To make that story short, again, one day I just decided, I think they're right. I had been arguing with them, and I got to just really dislike them because they were so nasty, and that we were stupid and they were smart. The attitude was awful. But I came to believe they were correct.[28]

According to Twachtmann, "We had to look up the word ecology in the late 1960s." The abandonment of the Fresh Water Lake project marked one of the many transitions that the district has made since its creation.

WATER WARS

Journalists often refer to water disputes in the Tampa Bay region as "water wars." Of course, calling political opponents a few names and filing some lawsuits is far from "war." This metaphorical shorthand for state water disputes is based on a larger myth about international water wars. Although it flies in the face of popular understanding, countries hardly ever go to war over water. One way or another they resolve water disputes peacefully, as has been the case in the Middle East.[29] Florida has had water conflicts, but not a war.

There certainly is a history of water supply conflict between the Southwest Florida Water Management District and the cities around Tampa Bay that withdraw water from wellfields to the north. There are limited supplies of surface and groundwater on the Pinellas peninsula, and St. Petersburg looked north for water as early as the 1920s. By the 1960s, most communities in the county depended on transporting water from adjoining Hillsborough and Pasco counties for much of their water supply.

The intergovernmental conflicts intensified as Tampa Bay became more densely populated. Only two years after the creation of the Water Resources Act of 1972, the legislature amended it by authorizing local governments to form intergovernmental entities called regional water supply authorities. The West Coast Regional Water Supply Authority, in Hillsborough, Pasco, and Pinellas Counties, was the first to form. The district and the Authority were at odds over safe levels of withdrawals from the wellfields operated by the Authority. The partial regionalization of water supply in the Tampa Bay area did not resolve the water conflict. In the 1970s the district responded in part by adopting a unique water resource management rule aimed at water resource sustainability. As a general guide in permitting, a water use permit applicant was limited by the rule to withdrawing no more than 1,000 gallons of groundwater per acre per day (Rule 16-J-2.11). As a result, operators of wellfields that pumped large volumes of water would have to acquire large tracts of land. Pinellas County and the West Coast Regional Water Supply Authority challenged the rule and persuaded the state Hearing Officer that the rule was "hydrologically unsound" and was not authorized by the Water Resources Act.[30] The rule was overturned in 1980.

The Southwest Florida Water Management District had to then find other means to prevent excessive water withdrawals. In the 1990s the Southwest District and the other districts began to adopt minimum flows and levels (MFLs) within the framework of the original 1972 Act. Today the Southwest District has as many adopted MFLs as any other district.

In the 1990s, as the Tampa Bay dispute continued, the Southwest District tried the method of monetary persuasion with the Regional Water Supply Authority and local governments. The struggles with water suppliers in the Tampa region had been going on for many, years when, in 1993, the district's main governing board and its basin boards set up what was called the New Water Sources Initiative. Prompted by Chair Joe Davis and his successor Roy Harrell, the district began to set aside $10 million a year in a fund to entice water utilities to develop new alternative water supplies.[31] It was a good deal for local governments that accepted the money. The district put up 50 percent of the costs of water supply projects (25 percent from the main governing board and 25 percent from the basin boards). Farmers got an even better deal: using its Facilitating Resource Management Systems program, the district would pay a farmer up to 75 percent of the cost of farm equipment that would reduce withdrawals from the upper Floridan Aquifer (Section 40D-26.401, F.A.C.) All of the districts allocate some funds for water supply assistance, but the Southwest District spends much more than the others for this effort.

The biggest payment from the fund came in the late 1990s, when the district decided to give Tampa Bay Water $183 million as part of a deal to settle disputes about overpumping of wellfields. Antagonistic feelings on both sides had risen to boiling point. As with many other seemingly interminable legal battles, there came a point at which the parties were willing to negotiate. The West Coast Regional Water Supply Authority had new members, including Steve Seibert, a Pinellas County commissioner who would later become the secretary of the Florida Department of Community Affairs and the first executive director of the Century Commission for a Sustainable Florida. In 1997, the Authority gave him and others the direction to try to cut a deal with the Southwest Florida Water Management District.

After intense negotiations, aided by severe legislative pressure, in 1998 the West Coast Regional Water Supply Authority and the Southwest District signed a peace treaty they called a Partnership Agreement.[32] The Authority agreed to develop at least 85 mgd of new water projects, which would make up for reducing the overpumping at the wellfields north of the metropolitan area. The Authority also agreed to reduce pumping at the wellfields from 158 mgd to no more than 121 mgd by the end of 2002, and to no more than an average of 90 mgd by the end of 2007. The cutbacks would be partly achieved by increased efforts at water conservation, but mostly by developing new supplies. Building the new water supply sources proved to be difficult.

Although the Authority did meet both the 121-mgd and 90-mgd pumpage targets in less than a decade, both of the two new large water supply projects built by the Authority (reorganized and renamed as Tampa Bay Water) ran into serious problems. A seawater desalination plant that Tampa Bay Water built received international attention. The 25-mgd plant was designed to be the largest in the Western Hemisphere. Tampa Bay Water devised what seemed a low-risk way of building the plant: companies would bid on contracts to construct a plant and supply desalinated water at a fixed price per thousand gallons. The deal was acclaimed as a model public–private partnership and a way of making the region more drought-resistant.

The problem was that the private side of the partnership did not live up to its side of the bargain. A consortium of companies, under the name of S&W, LLC, signed the contract to supply desalinated water at the very low price of $1.71 per thousand gallons (with the assistance of $90 million from the Southwest District funneled through Tampa Bay Water). The consortium ran into such serious financial troubles that Tampa Bay Water had to take over full ownership of the plant in 2003. They contracted with another firm to run the plant they had unwillingly come to own. That deal also went sour, and another firm had to take over. Tampa Bay Water also discovered that another $29 million had to be spent to fix operational problems. In 2009, 6 years later than originally expected, the desalination plant finally produced 25 mgd of fresh water.

The second large water supply facility built with the assistance of Southwest District dollars also ran into difficulties. Tampa Bay Water built the state's largest aboveground water supply reservoir in southeast Hillsborough County, adjoining the Alafia River. In a time-honored tradition, the official name of the reservoir is the C. W. Bill Young Regional Reservoir, named after a local Congressman who helped secure a Congressional earmark for $57 million of the $146 million cost of the reservoir. The ribbon-cutting ceremony with the Congressional representative was held on October 10, 2005, when the reservoir was nearly full. The Congressman, running for reelection in 2008, asked his constituents to "Remember, every time you turn on your water faucet, the water that comes out just might be flowing from the C. W. Bill Young Reservoir that was funded by a Congressional earmark."[33] But for part of 2009, no faucets dispensed water from that reservoir – it was nearly empty. So many cracks had opened up on the inside of the reservoir levee that Tampa Bay Water had to empty it just when water was most needed in the drought of 2007–2009. Temporary repairs enabled the reservoir to refill with recent above-average rains, but permanent repairs are expected to cost more than another $100 million.

In 2007, when the $183 million Partnership Agreement was about to expire, the Southwest District signed up to give Tampa Bay Water another $116 million for further expansions of their water system.[34] The dollars from the district were supposed to help drive down the costs of delivering water to customers in the region. Tampa Bay Water was not the only water supply organization that received many millions of dollars from the district. For example, the Peace River Manasota Regional Water Supply Authority in the southern part of the Southwest District received $80 million from the agency for building its new $168 million off-line reservoir on the Peace River.

WATER CONSERVATION

A major part of the water supply answer for the Southwest District is improved water use efficiency. Earlier than other water management districts, Swiftmud recognized the potential for improving water conservation. Of the five water management districts, only the Southwest District has a mandatory rule limiting the per capita use of water by urban suppliers. The cap is 150 gallons per capita per day (gpcd) but it allows up to a decade for communities above that threshold to achieve that level of efficiency. With the per capita rule and other measures that the district implemented or funded, water use per person is lower in the Southwest Florida Water Management District than in any other district.

However, to the extent that the district's subsidy of water supply development makes water cheaper, and thereby encourages demand for water, the Southwest District is working at cross-purposes. At one end of the water supply process it pays half of the cost of developing new supplies, which makes it cheaper to produce and sell water. However, because customers use more water when it is cheaper, water conservation becomes more necessary. The district helps fund water conservation projects, too – in part to reduce the demand for water that their water supply subsidies helped create.

A SUSTAINABILITY TRANSFORMATION?

The Southwest Florida Water Management District has undergone a series of remarkable transformations. Only a few years after its beginning, it turned away from an effort to convert part of Tampa Bay into a freshwater lake. In the 1970s the district decided that it must revamp the Four River Basins Project designed by the Army Corps of Engineers, even though that meant turning away from many millions of dollars in federal appropriations. By 1985 the chair of the district's governing board could reflect on how narrowly they had escaped from following the pattern of old-style Florida water projects:

> We are going to try to set the tone for the future here. We were awfully fortunate we didn't embark on a structural solution or we'd be faced with the same problems that plague the Kissimmee River.[35]

Although the district managed to not channelize or impound as many water resources as contemplated originally, it nonetheless was confronted in the 1970s with water supply controversies far beyond those imagined in 1960. The 15 years the district devoted to devising an implementation strategy and an enforceable rule for the Southern Water Use Caution Area demonstrate its staying power.

It is possible that the Southwest Florida Water Management District may lead Florida water management into an era of water resource sustainability. However, the challenges to sustainable water management in the district will require more dramatic adaptations than any it has previously accomplished.

MOVING WATER CONSERVATION TO THE NEXT LEVEL

Although the Southwest Florida Water Management District is a recognized leader in water conservation, it nonetheless has enormous opportunities to promote water use efficiency. Reducing landscape irrigation in the region leads the list. Due to below-average rainfall in 2007–2009, the district restricted landscape irrigation to only one day a week, which was at first regarded as draconian. However, the region seems to have survived very well. Many residents have changed their landscaping to more sustainable plantings, installed more efficient irrigation systems, ceased irrigation altogether, or turned to reclaimed wastewater for irrigation. (After the drought eased, the district's mid-2010 decision to allow 2 days a week of landscape irrigation appears to contradict its water conservation message.)

By the power of example, the district could also lead Florida in other sectors of water use efficiency. It could, for instance, issue a water use permit for public water supply only if a local government applicant agrees to allow only U.S. Environmental Protection Agency (EPA) WaterSense-certified plumbing fixtures to be installed in the community. Or it could issue permits only to public water suppliers that firmly restrict the installation of water-thirsty new landscapes. An obvious source for funding of water conservation is the money currently allocated for water supply development projects. If the hundreds of millions of dollars now going to water supply augmentation were diverted to water conservation, there would be less need for the supply projects. It makes little sense to subsidize both the construction of water supply projects and water conservation projects. The district allocated $127.7 million for cost-share projects in the fiscal year 2009–2010 (43 percent of the entire budget).[36] Moving most of this into water demand management projects would move water conservation to the next level.

A second possible source of funding for water conservation projects is raising the fee for water use permit applications (or permit monitoring) to a level closer to the cost of processing them. Water users in Swiftmud withdraw more than 1.1 billion gallons a day of water but the agency collects only $3 million a year for permitted activities of all kinds – less than 1 percent of its revenues.[37]

A third major possible funding source for water efficiency projects would be for the Florida legislature to adopt a system of water use fees, wherein water users that withdraw more water would pay a fee proportionate to that volume. (This is described more fully in Chapter 12, Charge a Few Nickels.) A water use fee would reduce water use to some degree, due to the price signal of paying for excessive consumption. The revenues from a fee might be used to fund water efficiency programs or partially reduce current property tax assessments.

LEADING ON WATER RESOURCE SUSTAINABILITY AND CLIMATE CHANGE ISSUES

The planetary climate change now under way challenges the sustainability of the Southwest Florida Water Management District as much as anywhere else in Florida (as explained in Chapter 14, Turning Up the Thermostat in Paradise). Some current Swiftmud activities do not help with climate change mitigation and adaptation goals. For example, the Alafia River and Peace River water supply projects involve high energy demands both for construction and operation. The seawater desalination plant that the district helped pay for produces up to 25 mgd of water, but with an energy cost far higher than that of other water production facilities. Meeting the challenge of climate change in Florida will mean dramatic reductions, rather than increases, in the emission of greenhouse gases. Water supply development in the future should focus on low-energy and low-carbon alternatives rather than capital-intensive construction projects that require high energy inputs for every gallon of water.

Perhaps the unique basin board system of the Southwest Florida Water Management District may help the district in meeting twenty-first-century sustainability challenges. The basin boards were designed from the start to ensure that the district was responsive to local concerns. Global climate change, however, is also meaningful for regions as small as this district and its even smaller basin boards. Droughts, floods, and warmer temperatures are global phenomena but occur in smaller Southwest District communities and watersheds, too. If the Southwest Florida Water Management District rises to these new sustainability challenges, it will have reinvented itself once again to meet contemporary needs. Alfred McKethan's vision of a truly comprehensive water *management* agency will have been vindicated again in the new century.

NOTES

1. U.S. Army Corps of Engineers. (1962) "Letter from Secretary of Army to Congress on the Four River Basins, Florida," House Document 585, September 26, 1962

2. Wakefield, John W. (1960) "Statement to the Public Works Subcommittee of the Appropriations Committee of the United States Senate on behalf of the request of the state of Florida for a reconnaissance study of flood control and water conservation needs of the Withlacoochee River, the Oklawaha River, the Peace River, the Hillsborough River, and adjacent watersheds in west central Florida," April 13, 1960

3. Alfred McKethan is given substantial credit for the creation of the Southwest Florida Water Management District, but he said that the idea for the district came from two Hernando County politicians: state senator James E. Connor and state representative John L. Ayers. Sam Mase, "New water district project is started," *St. Petersburg Times*, 29 April 1964

4. Wakefield, John W. (1960) "Memorandum to delegates to the Central and Southwest Florida Water Management District Steering Committee, boards of county commissioners with the proposed district, and interested individuals," September 2, 1960

5. Bryant, Farris [Florida Governor]. (1961) "Remarks at organizational meeting of the Governing Board of the Southwest Florida Water Management District, Tallahassee, Florida, August 28, 1961"

6. Oral History Program. (2006) Dale Twachtmann, 20 February

7. Southwest Florida Water Management District. (1967) *SWFWMD and the Ark*, Southwest Florida Water Management District, Tallahassee, FL

8. Heyl, Michael G. (2008) *Weeki Wachee River Recommended Minimum Flows and Levels*, Southwest Florida Water Management District, Tallahassee, FL

9. Southwest Florida Water Management District. (2001) *Springs Coast Comprehensive Watershed Management Plan*, Southwest Florida Water Management District, Tallahassee, FL

10. Southwest Florida Water Management District. (2008) *Surface Water Improvement and Management Program: 2007 Annual Report*, Southwest Florida Water Management District, Tallahassee, FL

11. Flannery, Michael S., et al. (2010) *The Determination of Minimum Flows for the Lower Myakka River* [peer review draft], Southwest Florida Water Management District, Tallahassee, FL, pp xxxviii and 8−27 + .

12. Townsend, Billy. (2007) Quoting Jeff Spence in "Peace River restoration means loss of homes," *Tampa Tribune*, 27 November

13. Southwest Florida Water Management District. (2009) "Peace River: The discovery of phosphate," www.swfwmd.state.fl.us/education/interactive/peaceriver/phosphate.php, accessed 27 December 2009

14. Florida Geological Survey. (2005) *Florida Geology Forum*, Florida Geological Survey, Tallahassee, FL, p7

15. Rockström, Johan, Steffen, W., Noone, K., Persson,Å., Chapin, III, F. S., Lambin, E., et al. (2009) "A safe operating space for humanity," *Nature*, vol. 461, pp472−475

16. Haley, Sandra. (1996) "Less is more," in *Florida Water*, vol. 5, no. 2, p2

17. Lloyd, Karen. (1999) "Southern Water Use Caution Area II," *Environmental and Land Use Section Reporter*, Florida Bar, vol. XXI, no. 1

18. Oral History Program. (2004) Derrill McAteer, 22 June

19. Blount, Archie. (1985) "Legislators block water district move," *Tampa Tribune*, 21 June

20. Flinchbaugh, Patrice. (1984) "Graham sets deadline for end of water board feud," *St. Petersburg Times*, 15 November

21. "Flood control district: Right step," *St. Petersburg Times*, 20 April 1960

22. U.S. Army Corps of Engineers, op. cit. note 1

23. Oral History Program, op. cit. note 18

24. Oral History Program, op. cit. note 6

25. Morgan, Lucy Ware. (1972) "Vast changes proposed for Green Swamp plan," *St. Petersburg Times*, 17 July

26. Powell, Gary L. (2005) *Minimum Flows for the Tampa Bypass Canal*. Report to the Southwest Florida Water Management District, Tallahassee, FL

27. Sumner, Ralph. (1951) "Engineer seeks conversion of bay near Campbell Park into fresh water lake," *Tampa Morning Tribune*, 29 January, p5A

28. Oral History Program, op. cit. note 6

29. Barnaby, Wendy. (2009) "Do nations go to war over water?" *Nature*, vol. 458, pp282−283

30. Florida Division of Administrative Hearings. (1980) Case No. 79-2352R and 79-2393R, April 9, 1980

31. Rand, Honey. (2003) "Water wars: A story of people, politics, and power," *Xlibris*, 2003, p127

32. The formal name of the partnership agreement is the Northern Tampa Bay New Water Supply and Ground Water Withdrawal Reduction Agreement between West Coast Regional Water Supply Authority, Hillsborough County, Pasco County, Pinellas County, City of Tampa, City of St. Petersburg, City of New Port Richey, and Southwest Florida Water Management District. The agreement, dated April 20, 1998, is 45 pages long, plus 173 pages of attachments.

33. Troxler, Howard. (2008) "Buying our favor with our own debt," *St. Petersburg Times*, 15 March

34. *Tampa Bay Water Financial Statements for Tampa Bay Water, Years Ended September 30, 2008 and 2007; Tampa Bay Water; and Tampa Bay Water April 20, 2009*, Board agenda package.

35. Gallagher, Pete B. (1985) "The Green Swamp may benefit from past environmental mistakes," *St. Petersburg Times*, 23 January, quoting Bruce Samson, the chair of the SWFWMD Governing Board.

36. Southwest Florida Water Management District. (2009) *Fiscal Year 2010 Annual Service Budget*, Southwest Florida Water Management District, Tallahassee, FL, p12

37. Ibid, p21

CHAPTER EIGHT

The St. Johns River District: Reaching the Limits of Water Supply?

T he 310-mile length of the St. Johns River makes it the longest stream entirely within Florida and the longest north-flowing river in the United States. The St. Johns is one of only 14 rivers designated by the U.S. Environmental Protection Agency as an American Heritage river, and in 2008 it was named by the American Rivers organization as one of the "ten most endangered rivers" in the United States. Water supply recently became the biggest issue for the St. Johns River Water Management District. Fragile water resources in St. Johns are already stressed from overwithdrawals, but water utilities nonetheless are projecting huge increases in the demand for water in this fragile region.

The upstream (southern) end of the St. Johns River is an immense complex of wetlands through which water slowly flows to the north and eventually makes a discernible river channel. The middle and lower sections of the St. Johns include Lake George (the second largest lake in Florida), major tributary streams, a 50-mile-long stretch of river averaging 2 miles wide, and a final passage through metropolitan Jacksonville before emptying into the Atlantic Ocean. The entire St. Johns River watershed is about 8,700 square miles in area, reaching from vast marshes far upstream to the Atlantic Ocean. However, even that large watershed comprises only about 70 percent of the 12,400 square miles encompassed by the St. Johns River Water Management District. The district includes watersheds that adjoin the large river or are to the north of Jacksonville at the mouth. The district is also charged with managing one-third of the lakes in Florida and more than 50 springs. St. Johns River tributaries like the Ocklawaha and Econlockhatchee themselves are large river-and-lake systems with their own history of human abuse and consequent need for repair.

Like most of Florida, the St. Johns River is geologically young. One hundred thousand years ago it was not a river but, instead, an estuarine lagoon. A series of barrier islands that once were on the eastern side of the lagoon became connected to each other after sea levels dropped, and they now separate the St. Johns River from the Atlantic Ocean. The bottom of the lagoon was naturally quite flat and so is the St. Johns River today. It flows northward, very slowly. The prehistoric estuary has a nearby successor in another ribbon-like estuary in the district – the Indian River Lagoon. The lagoon (a National Estuary) parallels the river and the Atlantic Ocean along much of the St. Johns, running 156 miles from Volusia to Palm Beach Counties. The lagoon is one of the most diverse estuaries in North America, supporting more than 4,300 species of plants and animals. Unfortunately, almost of a fifth of the Indian River Lagoon's seagrass meadows have been lost to development since 1943.[1]

Some of the largest springs in the state, including Volusia Blue, Silver Springs, Alexander Springs, and Salt Springs, make a contribution to the river. The groundwater resources of the district are large enough to be the primary source of drinking water for 97 percent of its residents and to supply many large springs. However, so much water is taken from the aquifer that a major fraction of the district cannot safely withstand additional withdrawals.

The St. Johns River Water Management District must manage the extraordinary water resources of one-sixth of the state, including its eponymous river. To carry out its diverse responsibilities, the district has acquired almost 700,000 acres of land – more than any other district except the South Florida Water Management District. The district has an annual budget in the range of $300 to $400 million, supported by a staff of 700. Like the other water management districts, the number of employees has not been allowed by its governing board to increase for at least a decade. Because more people have moved to Florida over that period, the number of employees per district resident in the St. Johns district has declined by about 20 percent over that period.[2] Meeting water resource challenges with more residents but no additional employees is becoming more difficult.

Three problems dominate the history of this water management district. One is the need to restore the environmental functions of the highly altered Upper St. Johns River. A second need is to reverse the many kinds of damage done in the even more complex Ocklawaha River and lake system. Both of these restoration efforts were opposed by some local interests but nonetheless were broadly popular and received substantial legislative support. The third big issue is more recent and even more challenging: rapidly increasing demands for drinking water from cities.

BEGINNINGS OF THE ST. JOHNS RIVER WATER MANAGEMENT DISTRICT

Like the other Florida water management districts created in 1972, St. Johns had a full agenda of water problems when it began operations. Art Marshall was named as the first governing board chair at its first meeting in Deland on November 28, 1973. Marshall became much better known later in his life as a prominent advocate of Everglades restoration and a strong ally of Marjory Stoneman Douglas in that campaign. He served as chair only until 1975 but was remembered by Tommy

Clay, the next chair, as a "very, very capable man" who also "openly bucked the governor on some policies."[3] For example, Marshall strongly opposed Governor Reuben Askew's 1975 proposal to grant the water management districts the power to levy an ad valorem property tax. He promised that "I will do everything in my power" to see that the proposal – one of Governor Askew's top priorities – fail.[4] Marshall was not reappointed to the board that year.

Tommy Clay, a Putnam County rancher and USDA county extension agent, became the next chair of the district. Like Alfred McKethan in his role as the first chair of the Southwest Florida Water Management District, Clay wanted the headquarters of the new agency placed in his home county. At one point during his term as chair, Clay even resigned temporarily to protest a study of putting the headquarters anywhere other than in Palatka. He said that he "had a great hand in getting the district located in Palatka" (the seat of Putnam County). The district constructed a headquarters building in Palatka and named it after him in 1980.

The Upper St. Johns River and its environmental degradation were top priorities for the new district, but there was a crippling limitation: much of the upper watershed was not included in the original legal boundaries of the district. In 1973, the Central and Southern Florida Flood Control District (predecessor to the South Florida Water Management District) reached from extreme south Florida on the Atlantic Coast all the way north to the middle of the headwaters of the St. Johns River. Both Marshall and Clay made it a priority to extend the original boundaries of the new St. Johns district southward to pick up the entire headwaters.

Some in the South Florida district resisted transferring the St. Johns River headwaters to the new district. For one thing, until the 1976 Florida constitutional referendum, the St. Johns district did not have dependable financial resources to operate and maintain the structural works already in place in the Upper St. Johns, built by the C&SF Flood Control District and the U.S. Army Corps of Engineers.[5] The struggle caused lingering resentments. Jack Maloy, with the South Florida Water Management District, claimed that the St. Johns board "looked at us like a cow looking at last year's calf."[6] Nonetheless, the St. Johns District did become custodian of the entire headwaters on January 1, 1977.

Challenge One: The Upper St. Johns Project

The district's water resources were damaged in the state's long quest to dike, drain, fill, channelize, subdivide, and farm wet areas. Many years before efforts to restore the Everglades in south Florida received worldwide media attention, the St. Johns District commenced restoring the headwaters of the St. Johns River. In some ways, that relatively unknown effort is a model for current Everglades activities. More than one-third of the upper basin (at the south end of the river) is wetlands. In the enormous marshes, cypress swamps, and hardwood swamps in Okeechobee County west of Vero Beach, hydrologic boundaries are indistinct. On the eastern side of the expansive wetlands, sheet flows of water go north and eventually form a distinct channel for the St. Johns River. On the western side, the marshes contribute flow to the Kissimmee River, which flows southward into Lake

Okeechobee. The southern sections of this vast wetland flow toward the St. Lucie River. Because the headwaters have so little topographic relief, the St. Johns River has no distinct river channel for its first 30 miles of flow northward. At least, those were the natural directions of the diffuse flows in the watershed before steam shovels and dredges began their work.

These large machines were deployed in the Upper St. Johns in the late nineteenth century and continued to operate on different projects in the watershed for a century. Private parties and single-purpose drainage districts undertook the work. The primary purpose, like most such Florida projects, was agricultural: making wet organic soils suitable for row crops, beef cattle, and citrus groves.[7] Much of the original 400,000 acres of wetlands was drained by the middle of the twentieth century. The Central and Southern Florida Flood Control Project of 1948 also helped change the landscape. Although the C&SF Project originally included no specific drainage works in the Upper St. Johns, Congress amended the project in 1954. Reengineering for flood control and drainage began actively in 1962, the very same year that the Corps began channelizing the Kissimmee River 25 miles to the west. The main engineering ideas were to channelize the marshy floodplain of the St. Johns, send floodwater east to the Indian River Lagoon, and store runoff in a vast series of low-level reservoirs along the ridge, west of the river. John McPhee reported on one of the most massive wetland conversion projects undertaken by a private corporation:

> In 1959, the Minute Maid Company went into the savannas with earth-moving machines and heaved up a great ten-foot wall of earth surrounding seven thousand acres of marsh. Then they pumped out the water, graded the sandy soil, and planted six hundred thousand orange trees. It was an impressive feat, and it emboldened many other companies and syndicates to do the same. The water is pumped over the dike into state flood-control canals.[8]

By 1970 the Upper St. Johns Project's main canal to the Indian River Lagoon was operational and the initial system of reservoirs in the Upper St. Johns was nearly complete. The canal began draining large areas of wetlands within the basin and diverting flood flows to the Indian River Lagoon.[9] By 1972, 62 percent of the marsh had been drained. The canals, by deliberate design, sent water rushing to the Indian River Lagoon in damaging flood quantities. The rest of the project plans would eliminate even more wetlands, including the Jane Green reservoir, which would flood and destroy thousands of acres of forested wetlands.

However, public opinion and new laws were turning against these kinds of land conversions. Farms in the upper watershed were releasing large amounts of nutrient pollution to the Indian River Lagoon through the new canals. Valuable fish and wildlife populations of the marshes and lagoon were shrinking. A new federal law requiring the Corps of Engineers to write an Environmental Impact Statement for this kind of activity played a key role in stopping the project. The Environmental Impact Statement found conditions alarming enough that both the Corps and the water management district decided to suspend project construction in 1972.

The massive watershed problem was presented to the brand-new St. Johns River Water Management District, which had no permanent headquarters, no independent source of revenue, and only the beginnings of a staff. Nonetheless, the Upper St. Johns River became the highest district priority in its early years. After consultation with a technical advisory committee, the district recommended substantial redesign of the outdated Corps project. The new design in the late 1970s kept the old mission of flood control but attempted to reduce the associated environmental damage. Although the redesign would rely upon many levees and structures, it would be only "semistructural" and would attempt to imitate the natural hydrology of the Upper St. Johns River.[10] A set of water management areas and marsh conservation areas would amount to more than 165,000 acres in Indian River and Brevard Counties. Project managers, after making enormous land purchases, would try to rearrange the altered hydrology of these areas to mimic the natural cycles of high and low water. If successful, this could help recreate the healthy ecosystem that once thrived in the Upper St. Johns. Water quality would be improved, as would fish and wildlife habitat, beyond the narrow purposes in the Congressionally authorized Corps project.

The water management district's governing board unanimously endorsed the redesign in 1983 and sent it up the Corps hierarchy for consideration. It took three years for the Corps to approve the new design, but the changes were judged to be so large that they could not be carried out under the terms of the project originally authorized by Congress. It was not until 1988 that Congress legally authorized the changes so the Corps could finally begin to build the new version of the project.

Both the district and the Army Corps of Engineers now hail the revised Upper St. Johns Project as a "model of modern floodplain management – balancing the needs of a river with those of the people and creatures who depend on it."[11] In 2008, the Upper St. Johns Project won an international prize, commended as "one of the largest river restoration projects in the United States and[. . .]a 30-year collaboration between state and federal water managers."[12] The prize was deserved. Environmental restoration was a relatively new and untested idea in the 1970s and 1980s when the St. Johns River Water Management District tackled the Upper St. Johns Project. The district and the Corps have improved or restored tens of thousands of acres of wetlands. Imperiled birds like the snail kite and the American wood stork nest within project boundaries. Water quality in the Indian River Lagoon has also improved significantly.

However, the restoration project does have major limitations. It is still a flood control project and is only "semistructural" rather than a fully self-maintaining natural hydrologic system. Much of the enhanced water resource features are confined behind managed levees and canals, with names like Fellsmere Area 1, Levee L-74N flow-way, and Structure S-161.[13] The project provides subsidized water supply and other benefits to farms in the upper basin, like those contemplated in the 1954 version of the project. The Corps' "operating manual for the Upper St. Johns Project spells out the agricultural water supply purpose:

> The primary purposes of the Upper St. Johns River Basin project are to provide flood control, provide environmental benefits by improving

environmental conditions in the marsh, and to enhance agricultural water supply potential within the basin. The objectives of the Interim Plans for water control are to permit construction while providing flood protection, preserving the surrounding marsh ecosystem, and maintaining agricultural water supply.[14]

The St. Johns River Water Management District is also clear about another agricultural purpose for the project. Elements of the constructed project are intended to receive polluted runoff from farm properties:

> Water from nearby citrus groves and livestock pastures is now discharged into large reservoirs called water management areas, which keep agricultural runoff isolated from water that recharges the marsh. This runoff water can be reused for farm irrigation and freeze protection.[15]

Agricultural runoff is laden with farm chemicals and often does not meet state water quality standards. The citrus and other agricultural operations in the watershed are, in effect, using the constructed water management areas of the Upper St. Johns project for both water quality treatment and water supply. Even with this benefit afforded to agriculture, the Upper River still has water quality problems due to excessive agricultural nutrients. The Florida Department of Environmental Protection (FDEP) determined that the Upper St. Johns should be included in the statewide list of water bodies impaired by pollutants. The FDEP found that agriculture was a major cause of the water quality impairment because it constitutes 43 percent of the land use in the Upper River but contributes 73.4 percent of the total phosphorus loadings that help cause dissolved oxygen problems.[16] (The same trend is true in other river segments downstream. From Palatka to the city of Orange Park, 70 percent of the phosphorus entering the river comes from agriculture.)[17]

Farm interests zealously protect the benefits that the Upper St. Johns Project gives them. The Taylor Creek Reservoir in Osceola County, for example, was built as a component of the Upper St. Johns Project in the late 1960s. It is located on and provides water supply benefits to a 450-square-mile agricultural operation, Deseret Farms. In 2010, nearby cities and the water management district were interested in raising the reservoir level to make water available for water utilities. However, Deseret Farms strongly opposes sharing the water with the cities unless the cities pay for the privilege (even though the reservoir was constructed by the U.S. Army Corps of Engineers).[18]

Challenge Two: Ocklawaha River and Lakes

The Ocklawaha River system became famous in the late nineteenth century. The St. Johns River was Florida's first highway for northern tourists and the beautiful Ocklawaha was the main attraction on the route. In the late 1800s, tourists booked tours on steamboats that paddle-wheeled up the narrow and zigzagging Ocklawaha, overhung with a picturesque tree canopy and teeming with exotic Florida wildlife. The trip often culminated at Silver Springs, which was a wonder in its own right. Prominent Americans such as Ulysses Grant, Grover Cleveland, and Thomas Edison all took the winding trip. Many tourists used the abundant wildlife for target practice. A newspaper reporter saw the effects: "By their insane shooting at everything, the tourists were driving all birds, alligators, and animals from this portion of the river."[19]

Eventually, steamboat captains forbade shooting practice but other human alterations of the Ocklawaha accelerated. River tourists began to observe large rafts of cypress logs floating down to the St. Johns River and the large lumber mill in Palatka. Over the course of a century, the river and lake system was damaged even more than the Upper St. Johns River, but by the same methods. Portions of the Ocklawaha watershed were drained to create farms, as was done in the Upper St. Johns. Floridians also channelized the Ocklawaha system for navigation, lowered its lake levels for residential development, and dumped industrial and urban waste into its waters. The very resources that first attracted visitors were damaged by human abuse.

The Ocklawaha basin is relatively large at 2,769 square miles, which makes it the largest tributary to the St. Johns River. The *upper* (southern) Ocklawaha is a series of wetlands and chains of lakes, including Lake Griffin, while the *lower* (northern) Ocklawaha is the section of the watershed that flows out of Lake Griffin as the Oklawaha River on its way to the St. Johns. The extreme upper (western) portion of the Ocklawaha basin drains the northern part of the Green Swamp (shared with the Southwest Florida Water Management District.) Like the marshes of the Upper St. Johns, the Green Swamp is the headwaters of several rivers. From south to north, Orange Creek, the Harris chain of lakes, the Clermont chain of lakes and Palatlakaha River, and Lake Apopka are the largest water resource features. All of the major lakes and rivers were drained and channelized in the nineteenth and twentieth centuries.

The fate of Lake Apopka, the state's fourth largest lake, is another example of human alterations of Florida's water resources. In 1893, promoters completed the dredging of a major canal from the northern end of Lake Apopka to Lake Beauclair. A canal system now ran between all of the large lakes of the Harris chain of lakes. In building the Apopka-Beauclair Canal, the dredges lowered the level of Apopka by almost 4 feet. That made the marshes on the north side of Apopka marginally dry enough for farming, once they were further drained by a network of additional smaller canals. Some farming was conducted there until World War I but the area was still too wet.[20]

Further assaults on the 50,000-acre lake's biological systems followed World War I. The town of Winter Garden began dumping partly treated sewage to Lake Apopka in the 1920s. Citrus processing plants also used the lake as a receptacle for their wastes. A navigation lock and dam constructed in 1925 at Moss Bluff also helped regulate water levels in Lake Griffin and the Ocklawaha River. Few understood at the time that the health of the lakes of central Florida depended on cyclic conditions of naturally high and low water levels and on reducing the pollutants dumped into them.

Another milestone of harm for Apopka was crossed in the early 1940s. At the behest of local farmers, the legislature set up the Zellwood Drainage and Water

Control District as a quasi-governmental agency on the north shore. The Zellwood district and other private agricultural landowners amputated close to 20,000 acres of marshes from the northern shore of Lake Apopka, far beyond what had been tried before World War I. The new "land" of the north shore was so low that massive pumping was needed to make the muck soil dry enough for growing crops. In dry times, water for irrigation at the industrial scale was easy to obtain by seepage and by simple gravity flow out of Lake Apopka. When there was too much water in the farmlands, pumps moved water back into the lake. However, farm water placed back in the lake was loaded with farm nutrients and pesticides. At first, the nutrients seemed to not harm the lake; they even seemed to improve the sport fishery. Over time, however, excess nutrients stimulated excessive plant growth in the lake.

Lake Apopka was approaching biological catastrophe. Apopka was a famous sport fishing site in the 1940s, with 19 fish camps. It was called the "Bass Capital of the World." However, in 1947, the lake began to change dramatically for the worse. The thriving aquatic community of rooted plants stretching across much of the lake gave way to persistent algal blooms in the sunny Florida climate. Within a decade, the famous sport fishery collapsed. Native gizzard shad could feed on the dense algae populations and rapidly became the dominant fish species in Apopka, but they were of no interest to anglers. By the 1960s, Apopka was known as one of the state's most polluted lakes. Fish kills became a common occurrence in the pea-green lake. The bottom was covered with a fluid mud that was easily stirred up in storms and that smothered any aquatic plants trying to grow in the mucky lake bottom.[21]

The collapse of Lake Apopka did not matter to the farms on the north shore. They still could withdraw all the water they needed from the lake and then return their farm runoff to the lake at will. By the 1980s, Zellwood farms were placing 20 billion gallons of high-nutrient farm water in Lake Apopka every year, amounting to one-third of the volume of the lake.[22] More than 85 percent of the surface water entering the lake came from a total of only 19 north shore farming operations.[23] Although the Zellwood farms were profitable, their continued operation would ultimately have been self-destructive. The organic soil of the north shore farms was disappearing through oxidization and subsidence, just like the soil of the Everglades Agricultural Area on the south side of Lake Okeechobee today. The largest farmer, Duda & Sons, Inc., formulated a novel response to the loss of soil that their operations caused: it sought a federal tax break. In the 1970s, the enterprise sued the federal government to claim a depreciation deduction for the disappearing soil. They managed to get a federal district court to rule in their favor but lost on appeal. If Duda had prevailed, federal tax policy would have rewarded farm practices that maximized the loss of soil.[24]

Lake Apopka and most of the Ocklawaha basin was in ecological trouble because they had been abused for a century. In the 1970s local interests and the state Game and Fresh Water Fish Commission advanced the idea of drawing down the water of Lake Apopka. The state Department of Environmental Regulation organized a study on whether Lake Apopka could be fixed by dramatically lowering its level for a year or two. If the fluid mud at the bottom of the lake could be exposed to the sun and dried out for several months, it might compact into more normal soil and even stay that way when reflooded. The dried-out lake bottom possibly would become a proper substrate for desirable rooted aquatic vegetation. However, the $200 million state plan did not make much progress. For one thing, the only way to temporarily get rid of the water in Lake Apopka was to pump it downstream into the Harris chain of lakes. Downstream lake residents strongly opposed the idea of putting polluted Apopka water in their local water bodies. Local citrus growers on the south side of the lake also opposed the state plan. They benefited from the "lake effect" of having a massive body of warm water in the winter sheltering their groves against the threat of occasional winter freezes. The state plan faded away.

The St. Johns River Water Management District, local governments, state legislators, and private groups led the next era of restoration. The St. Johns District became the coordinator of a multipronged effort to improve conditions in Lake Apopka. In 1985 the legislature provided impetus for restoration by enacting the Lake Apopka Restoration Act, including a Restoration Council and Technical Advisory Committee staffed by the St Johns District. Further action was mandated by the 1987 legislature when it named Lake Apopka as one of the water bodies to be addressed in a new statewide Surface Water Improvement and Management Program.

The district took vigorous action to reduce nutrient pollutant loadings to Apopka. After receiving legal permission from the state Department of Environmental Regulation, they began regulating the quality of water discharged from the remaining north shore farms in 1988. However, land acquisition activities overtook their regulatory approach to water quality protection. In 1996 and 1997, the legislature appropriated $65 million for the district to acquire north shore farm land. The USDA joined in the effort and contributed more than $25 million for the same purpose. From all sources, more than $100 million was spent to acquire all of the farms on the north shore, thereby leapfrogging over regulatory difficulties. The Lake Apopka Restoration Area now totals 19,000 acres, including 9,000 acres acquired with USDA assistance. Farm lands are being converted back to wet areas and water treatment flow-ways.

Serious mishaps have occurred along the path of restoration. For example, the north shore farms became a site of a massive bird kill. The accident began with a well-intentioned plan to begin restoring farm land that the district had just acquired. In the fall of 1998, the district implemented an interim plan to keep water on farm land deep enough to prevent the growth of terrestrial vegetation. This management strategy also would prevent the usual autumn release of high phosphorus-laden water to the lake. Over the winter, the district intended to slowly lower the water levels in the farm land and commence construction on part of the marsh treatment area and flow-way. This comprehensive strategy had a danger, however: the possible release of pesticides from the submerged soil.

The district was well aware that the farmers had applied toxic pesticides to the land for decades.[25] A toxicologist advised the district that there was little or no risk of harm to wildlife. The district thought that they had identified and neutralized pesticide hotspots. As the fields remained flooded into the fall and winter, wading birds were attracted to the wet farmland and arrived in extraordinary numbers and unusual diversity. A few dozen white pelicans usually appeared on Lake Apopka in the farming years, but more than 3,500 came in December of 1998. Birders were very excited about the extraordinary viewing opportunities. The Audubon Christmas bird count of 174 species broke a 100-year record for inland sites in the United States.

But the extraordinary number of birds resulted in an extraordinary number of bird deaths. Large birds started falling out of the sky, poisoned by the organochlorine pesticides they ingested while foraging for fish in the abandoned farmlands. By the time the district managed to drain the area that was attracting so many birds, 441 white pelicans, 58 great blue herons, 43 wood storks, and 34 great egrets had died. The U.S. Fish and Wildlife Service concluded that the district had had sufficient information to anticipate the bird kills. There is irony in this fault-finding because it was not the district but the farms that had applied pesticides like dieldrin, toxaphene, and DDT for decades. Nonetheless, in 2003 the water management district accepted partial responsibility for the bird kills.[26]

The district intensified its effort to bury the polluted layer of soil and thereby prevent it from interacting with fish and wildlife. District contractors used the largest production tractors in the world to plow thousands of acres of former farm land 3 feet deep and invert the topmost polluted layer and the lower layer of cleaner soil.[27] The plan is that small fish can no longer feed in the contaminated sediments and transfer the pollutants up the food chain. The contractors also applied an alum compound to bind with the phosphorus in the soil that might be released by reflooding. Eventually, much of the restoration area will be turned back into healthy wetland habitat.

The district is taking other measures to reduce nutrient levels in highly polluted Lake Apopka. For example, it is harvesting the dense populations of gizzard shad. This removes phosphorus and nitrogen from the lake that is incorporated in fish bodies. There are ancillary benefits to shad removal. For example, the shad stir up the soupy lake bottom, thereby reducing any tendency it might have to solidify. That same foraging activity also deters the growth of desirable rooted plants in the lake. Shad waste also helps spur dense growth of unwanted algae. Removing shad addresses all of these concerns. From 1993 to 2007 the district removed about a million pounds of shad per year from the Harris chain of lakes, mostly from Lake Apopka.[28]

The troubled history of Lake Apopka, at the upper end of the Ocklawaha River and Lake System, demonstrates some of the ways in which Floridians injured the very water resources that make the state a special place. Other parts of the Ocklawaha also were channelized and polluted in the twentieth century:

- **Lake Griffin.** Lake Griffin, downstream from Apopka, lost 7,000 acres that were converted to farms and cattle pastures in the 1940s. Farmers also organized the construction of the Yale-Griffin Canal in the 1950s to further lower the lake level and dry out the marshes. Like the farm operations by Lake Apopka, the polluted farm water was pumped directly into Lake Griffin, with similar adverse effects on water quality and on native fish and wildlife. The water management district established a 1,500-acre treatment marsh on Lake Griffin in 1994, which has much improved the water quality. The district also has a program to remove gizzard shad from Lake Griffin.

- **Ocklawaha River.** In the early 1900s, with federal assistance, Floridians built a new navigation channel alongside the Ocklawaha and arranged for the abandonment and drainage of 15 miles of the natural Ocklawaha River channel and 15,000 acres of wetlands. The water management district, in the Sunnyhill and Ocklawaha Farms projects, is now re-establishing the type of wetland community that once existed on the Ocklawaha River.

- **Silver Springs.** Silver Springs creates the Silver River near Ocala. With an average flow of more than 500 million gallons a day, it makes up about half of the total flow of the Ocklawaha River. A recent look at 50-year-long trends in Silver Springs demonstrates serious harm over that period. Average spring flow has decreased about 20 percent, while the concentration of nitrate pollution has increased dramatically. The water is less clear and fish biomass has decreased by 96 percent.[29] It certainly does not help that Marion County, in which Silver Springs is located, has nearly 100,000 septic tanks placing 2.5 million pounds of nitrates in the aquifer each year.[30]

- **Cross-Florida Barge Canal.** The dream of digging a barge or ship canal across the middle of the Florida peninsula dates back to the 1820s.[31] During the administration of President Franklin Roosevelt, the Army Corps of Engineers commenced work on a sea-level canal across Florida, but this effort was suspended after only a year. The canal project suffered from national allegations that it was a product of pork-barrel politics. It also had substantial opposition within the state on the grounds that the canal would cut into the Floridan Aquifer and induce widespread saltwater intrusion. The editor of the *Florida Grower* charged in the 1930s that the canal would "be a greater calamity than any freeze or hurricane" the state had ever experienced.[32]

   After a three-decade interruption, the project was reconfigured and found new life in resumed federal appropriations. At the groundbreaking ceremony in 1964, President Lyndon Johnson observed, "God was good to this country [. . .] But in His wisdom the Creator left something for men to do for themselves."[33] The Corps moved rapidly to build the new version of the canal. By 1968, it had constructed an impoundment on the Ocklawaha River just upstream of where the Ocklawaha flowed into the St. Johns River and was ready to impound another stretch of the Ocklawaha. However, a new organization, Florida Defenders of the Environment, fought against further work on the canal. In 1971 President Richard Nixon halted work on the canal, marking it as the largest public works project ever stopped by environmental opposition. By 1976 the state's governor and cabinet requested that the canal project be deauthorized, which finally happened in 1991.

The impoundment on the Ocklawaha River still exists, despite vigorous efforts to remove it and thereby allow the Ocklawaha to run free. If it ever is removed, the St. Johns River Water Management District is the government agency that will have to issue the necessary regulatory permits.

Thus, from its headwaters to where it joins the St. Johns River, the streams, lakes, and wetlands of the Ocklawaha system bore the brunt of many years of short-sighted human actions. The good news is that with decades of effort and hundreds of millions of dollars, some of its natural resource values can be at least partly recovered. By purchasing and then carefully managing a total of about 40,000 acres in the Ocklawaha watershed, the St. Johns River Water Management District has improved water resource conditions. Water quality has improved in Lakes Apopka and Griffin. Large areas that once were part of lakes are being recreated as wetlands or are serving as stormwater treatment areas. Full restoration is not possible but the work done by the St. Johns River Water Management District demonstrates the benefits of having a water management agency that can gather large financial resources and focus on a problem for decades. Unfortunately, those same can-do attitudes are not always so helpful in solving water supply problems.

Challenge Three: Water Supply, but at What Price?

In the Ocklawaha Basin and the Upper St. Johns River, the St. Johns River Water Management District is trying to undo the natural resource damage done by previous generations of Floridians. Success will mean at least partial recovery of the water resources injured in the past. The mission is clear: bring back the environmental health of the region's lakes, wetlands, rivers, and estuaries. The district has been reasonably successful. Compared to environmental restoration, water supply has been a lower priority – until recently. Total water use in the district is no greater today than it was in 1975. The public water supply sector has increased, however, and may grow even larger.

For the first time in the history of the St. Johns River Water Management District, the region is facing difficult water supply demands from one end of the district to the other. Without additional water conservation, the St. Johns District forecasts that urban demand for water will increase 118 percent from 1995 to 2030 (even more than the forecast 106 percent increase in population.[34]) In northeast Florida, so much water is already being withdrawn from the Floridan Aquifer that it has reduced aquifer levels for scores of miles inland. Even springs in the adjoining Suwannee watershed appear to have had their flows reduced. If current water use patterns continue, the northeast region will be taking almost 50 percent more water out of the aquifer by 2030 than in 1995. The aquifer could decline by many feet in several northeast Florida counties, thereby inducing saltwater intrusion, lowering spring flows, drying up wetlands, and damaging shallow homeowner wells.

Water supply issues are multiplying. The following sections examine specific examples.

Central Florida Coordination Area. The St. Johns River Water Management District includes a portion of metropolitan Orlando. In central Florida and around metropolitan Orlando, the Floridan Aquifer is shared by the St. Johns River, the Southwest, and the South Florida Water Management Districts. Withdrawals in what the three districts have termed the Central Florida Coordination Area must be considered and managed as a hydrologic unit.

In 2007 the three districts jointly concluded that "sustainable quantities of groundwater in central Florida are insufficient on a regional basis to meet future demands."[35] The St. Johns District groundwater models forecasted that the minimum flows established for five springs would be violated if groundwater withdrawals continued to expand. Lake and wetland water levels (which have their own minimum levels) would also be violated if business continued as usual. As a result, the three water management districts separately approved a coordinated action plan for the Central Florida Coordination Area to cut back on potential increases in groundwater pumping. The key initiative was to put a cap on the amount of groundwater allowed to be pumped from the Floridan Aquifer.

This was a dramatic change in groundwater policy in central Florida. However, it cannot be called a surprise. As far back as 1986 the U.S. Geological Survey cautioned that groundwater withdrawals were increasing and may "eventually exceed the capacity of the system."[36]

Volusia Blue Spring: Water for Manatees or Cities? Volusia Blue Spring, located in a state park near DeLand, is the largest spring on the St. Johns River and also a major manatee sanctuary during the winter. The half-ton mammals cannot survive more than a few days in water temperatures below 68 °F. Coastal waters surrounding Florida often dip well below that in the winter. As a consequence, manatees habitually gather around freshwater springs during the cold season to enjoy steady water temperatures of between 71 and 73 °F. At Volusia Blue, the number of manatees is trending sharply upward. A total of 301 individual manatees used the spring at some time during the winter of 2008–2009.

If the district issued permits for too much groundwater pumping in the springshed, the flow of the spring would decline and cold river water would come farther up the spring run. There would be a smaller thermal refuge for manatees in the winter. Water utilities in the region are already very interested in withdrawing more water from the aquifer that feeds the spring, which may conflict with the minimum flow and level (MFL) rule for Volusia Blue adopted by the district. Will future St. Johns Governing Boards decide to protect manatees or to satisfy the demands of water utilities?

Desalting Ocean Water. The St. Johns District encourages water utilities to consider desalinating seawater. This might seem surprising in light of the delays, construction problems, and cost escalation of the Tampa Bay Water seawater desalination plant described in the previous chapter. However, the Tampa experience has not prevented northeast Florida water utilities from considering another desalination plant – especially if St. Johns will help pay for it like the Southwest District did for Tampa Bay Water.

The St. Johns River Water Management District may do so. The district helped draft the memorandum of agreement between several counties (Flagler, St. Johns, and Marion) and several cities (Palm Coast, Leesburg, and Deland) to design a seawater desalination plant for what is called the Coquina Coast region. The district has already set aside almost $15 million to help pay for a desalination plant. Local governments, with district and state support, are seeking even more financial assistance.

It is possible to build another desalination plant in Florida. The technology exists, even if costs may turn out to be higher than projected. However, it makes little sense to promote energy and greenhouse gas–intensive desalination plants at a time when the United States and other countries are resolving to cut greenhouse gas emissions dramatically. Moreover, district subsidies to build such a plant have the perverse effect of creating artificially cheap water. Making desalination cheaper to customers doesn't help when water conservation should be emphasized.

Water Bottling.   As described in Chapter 10, Water Supply Is Easy, the St. Johns River Water Management District overcame both strenuous local opposition and a lawsuit when they granted a permit in 2008 for a water bottling plant in Lake County. The aquifer permit was for only a half-million gallons a day but was issued in the midst of a drought. Many citizens and newspaper editorial writers inveighed against the district's decision to issue the permit. Although apparently consistent with the district's rules, and adequately protective of the resource, issuing the permit heightened public distrust of the agency. The district won the legal battle about the bottling permit but may have lost the public relations battle.

Water from the St. Johns and Ocklawaha Rivers?   The cities of central Florida get almost all of their drinking water from the Floridan Aquifer. Because the aquifer cannot withstand increasing withdrawals beyond those expected in 2013, the St. Johns District has strongly advised water utilities to seek alternative sources of water beyond that date, such as the Ocklawaha and St. Johns Rivers. How much surface water might be available? The district's answer is up to 262 million gallons a day (155 mgd from the St. Johns River and 107 mgd from the Ocklawaha River).[37] According to their planning level studies, removing 262 mgd would increase salinity downstream somewhat, but it would have only a minor effect on the St. Johns estuary. Some estuarine species in the mouth of the river would move upstream with the saltier water. In the district's estimation, these impacts would be acceptable.

In reaching this conclusion, the St. Johns River Water Management District expended more than $1.5 million for scientific and technical investigations to establish MFLs at six different points along the St. Johns River. The district believed they were on solid scientific ground and would be able to issue permits to take water out of the rivers. They were stunned to encounter vociferous opposition to the first such permit application, submitted by Seminole County near Orlando.

Seminole County proposed to remove 7.25 mgd from the middle St. Johns River. The water would be used for both drinking water and to supplement their use of reclaimed water in landscape irrigation systems. In early 2008, district staff brought their recommendation to the governing board: issue a 20-year permit for river water at a lower level of 5.35 million gallons a day. Strong opposition erupted in many communities downstream on the St. Johns River. Several local governments and a private organization called the St. Johns Riverkeeper filed a legal challenge to the proposed water use permit. In January 2009, an administrative law judge recommended that the district issue the permit because the application met regulatory criteria. He concluded, after a two-week hearing, that the withdrawal would have only a minimal effect on river salinity downstream.[38]

That did not satisfy opponents. On April 13, 2009 a raucous public meeting about the permit attracted nine newspaper reporters, staff from three television stations, and more attendees than any other meeting in the history of the district. More people showed up than would fit in the 279-person capacity of the governing board's meeting room. After several hours of testimony, in which only two people supported permit issuance, the governing board voted by a narrow majority of 5–4 to follow the recommendations of the administrative law judge. The permit issued ultimately allowed only 0.80 mgd of river water until 2014 and 5.35 mgd after that date. Governing Board Chair Susan Hughes called it "the most difficult decision that has come to the board in my term."[39]

In the face of so much concern about possible withdrawals, the St. Johns District decided to conduct additional studies and seek the opinion of outside experts. They contracted with independent scientific experts and the National Research Council (part of the National Academies of Science) to assess the effects of withdrawing more water from the St. Johns River. The study is expected to conclude in mid-2011 and may have a powerful effect on the feasibility of withdrawals from the St. Johns and Ocklawaha Rivers. The NRC's first assessment, released in August 2009, urged the district to not "rush to judgment" about the environmental effects of river withdrawal.[40] The fate of the proposals for removing 262 mgd from the river system is unclear at this time and will depend highly upon the final results of the NRC study. The scientific concerns certainly will take a high degree of effort, and substantial time, for the district to address. Fundamental policy issues may be harder to resolve.

A BUILT-IN CONFLICT OF INTEREST

Water supply battles on several fronts are a new and unpleasant experience for the St. Johns District. The district is in these conflicts because Florida water law guides the water management districts only part way toward sustainability. By statutory direction, water management districts are somehow to *supply all water demands* of Floridians while *also fully protecting water resources*. The Florida Water Resources Act of 1972 declares that it is state policy to "promote the availability of sufficient water for *all* [emphasis added] existing and future reasonable-beneficial uses and natural systems." (Section 373.016(3)(d), F.S.) The law imposes the same unrealistic goal on water management districts when they prepare 20-year regional water supply plans. The plans are required to identify sources of water that will "exceed the need" identified in the plan for "*all* existing and future reasonable-beneficial uses within the planning horizon." (Section 373.0361(2)(a), F.S.)

This cornucopian vision is unattainable, particularly when water resources are already under serious strain. As the NRC pointed out, this conflict between goals is intensifying in the St. Johns River Water Management District. All of the Florida water management districts are deeply involved today in promoting, planning, and paying for water supply. The ability to fund water supply projects points to an inherent conflict of interest in the modern statutory framework for water management districts: the districts both permit water withdrawals and help develop water supplies. This conflict of roles within the St. Johns River Water Management District makes Floridians doubt the district's impartiality about permitting withdrawals from the St. Johns and Ocklawaha Rivers. The district has been telling local governments for a decade to consider getting their future water there. Can the district reasonably be expected to turn away a permit application for which it previously extended a welcome mat?

The most obvious way to avoid a conflict between water supply permitting and environmental protection is to avoid the need for developing more water supplies: use water more efficiently. That allows the same amount of water to meet the needs of more water users. It also saves money because the cheapest available water usually is the water a utility already has. The St. Johns River Water Management District needs to promote water conservation with the same money, power, imagination, and determination as it has promoted water resource restoration in the Upper St. Johns and Ocklawaha watersheds. Although not yet a state or national leader in water use efficiency, St. Johns appears to be moving in that direction. For example, in 2009 St. Johns adopted a rule limiting landscape irrigation to 2 days a week most of the year and to only 1 day a week in the winter. The district also is considering other advanced water conservation policies. Without even more emphasis on water conservation, the St. Johns River Water Management District will be locked into a cycle of developing more and more water supplies to satisfy growing demands due to inefficient use of water.

NOTES

1. St. Johns River Water Management District. (2005) *District Water Management Plan*, p50

2. St. Johns River Water Management District. (2009) *2008 – Citizen's Report for Fiscal Year Ended September 30, 2008*

3. Oral History Program. (2004) Tommy Clay, 14 October

4. Long, Evelyn. (1975) "SJRWMD chief raps taxing plan," *Florida Times-Union*, 18 June

5. Shields, Harmon [executive director of the Florida Department of Natural Resources]. (1975) Letter to Warren Hyten, February 28, 1975

6. Oral History Program. (2003) Jack Maloy, 12 November

7. St. Johns River Water Management District. (2007) *The Upper St. Johns Basin Project: Giving Nature a Second Chance*, St. Johns River Water Management District, Palatka, FL

8. McPhee, John. (1966) *Oranges*, Farrar, Straus and Giroux, New York, p109

9. U.S. Environmental Protection Agency. (1992) "Restoration of Florida's upper St. Johns River basin helps heal headwater marshes," *Nonpoint Source News-Notes*, issue 25

10. Sterling, Maurice, and Charles A. Padera. (1998) "The Upper St. Johns River Basin Project: The environmental transformation of a public flood control project," Professional Paper SJ98-PP1, St. Johns River Water Management District, Palatka, FL

11. Op. cit. note 7

12. St. Johns River Water Management District. (2010) "Upper Basin Project wins prestigious Australian award," http://sjr.state.fl.us/Districtawards/index.html, accessed 21 March 2010

13. For a full picture of the different elements of the entire river, see Belleville, Bill. (2000) *River of Lakes: A Journey on Florida's St. Johns River,* University of Georgia Press, Athens, GA

14. U.S. Army Corps of Engineers. (2005) "Upper St. Johns Project," in *System Operating Manual, Central and Southern Florida Flood Control Project* [draft], vol. 6, p7−1

15. St. Johns River Water Management District, op. cit., note 11

16. Gao, Xueqing. (2006) "Nutrient and DO TMDLs for the St. Johns River above Lake Poinsett (WBID 2893L), Lake Hell n' Blazes (WBID 2893Q), and St. Johns River above Sawgrass Lake (WBID 2893X)," Tables 4.2 and 4.4, *Draft TMDL Report*, Florida Department of Environmental Protection, Tallahassee, FL

17. Way, Pam Livingston, Chad M. Hutchinson, and Alicia M. Steinmetz. (2008) "Development of a controlled release nitrogen best management practice for potato production in the tri-county agricultural area of northeast Florida," *Florida Watershed Journal*, vol. 1, no. 3, p12

18. U.S. District Court, Middle District of Florida. (2009) *St. Johns River Water Management District v. Farmland Reserve, Inc.*, Complaint for Declaratory Relief, Case No. 6109-CF-1103-ORL-28-DAB, filed June 25, 2009

19. Purdum, Elizabeth D. (2002) *Florida Waters: A Water Resources Manual*, Southwest Florida Water Management District, West Palm Beach, FL, pp20,21

20. Schelske, C. L., E. F. Lowe, L. E. Battoe, M. Brenner, M. F. Coveney, and W. F. Kenney. (2005) "Abrupt biological response to hydrologic and land-use changes in Lake Apopka, Florida (USA)," *Ambio*, vol. 34, no. 3, pp192−198

21. Although there is some debate about the exact causes of the lake's ecological collapse, there is no argument about the need for vigorous action to reverse the impairments. For an example of the debate, see Bachmann, R. W., M. V. Hoyer, S. B. Vinzon, and D. E. Canfield Jr. (2005) "The origin of the fluid mud layer in Lake Apopka, Florida," *Limnology and Oceanography*, vol. 50, no. 2, pp629−635.

22. Industrial Economics, Inc. (2004) *Final Lake Apopka Natural Resource Damage Assessment and Restoration Plan, Prepared for the U.S. Fish and Wildlife Service, with Assistance from the St. Johns River Water Management District*, St. Johns River Water Management District, Palatka, FL

23. St. Johns River Water Management District. (2008). *Lake Apopka: A Decade of Improvements Now Accelerating*, St. Johns River Water Management District, Palatka, FL

24. *A. Duda and Sons, Inc. v. U.S.*, 560 F. 2nd 669, 48 A.L.R. 775, October 7, 1977

25. Industrial Economics, Inc., op. cit. note 22

26. Memorandum of Understanding between the St. Johns River Water Management District and the United States of America, October 8, 2003

27. Grattan, Melora. (2008) "Monster plow helps turn the tide of tainted soil surrounding Lake Apopka," www.enviro-net.com, accessed 14 January 2010

28. Schaus, M. H. (2007) *Effects of Biomanipulation on Nutrient Cycles in Central Florida Lakes via Nutrient Excretion and Bioturbation by Gizzard Shad*, Project No. SK933AA, Final Report, St. Johns River Water Management District, Palatka, FL

29. Munch, Douglas A., David J. Toth, Ching-Tzu Huang, Jeffery B. Davis, Carlos M. Fortich, William L. Osburn, Edward J. Philips, Erin L. Quinlan, Michael S. Allen, Melissa

J. Woods, Patrick Cooney, Robert L. Knight, Ronald A. Clarke, and Scott L. Knight. (2006) *Fifty-Year Retrospective Study of the Ecology of Silver Springs, Florida*, Special Publication SJ2007-SP4, St. Johns River Water Management District, Palatka, FL

30. Cohen, Matthew J., Sanjay Lamsal, and Larry V. Kohrnak. (2007) *Sources, Transport and Transformations of Nitrate-N in the Florida Environment*, Special Publication SJ2007-SP10, St. Johns River Water Management District, Palatka, FL

31. Noll, Steven, and David Tegeder. (2009) *Ditch of Dreams: The Cross Florida Barge Canal and the Struggle for Florida's Future*, University Press of Florida, Gainesville, FL, pp13–15

32. Tegeder, Michael David. (2004) "Economic boon or political boondoggle? Florida's Atlantic Gulf Ship Canal in the 1930s," *Florida Historical Quarterly*, vol. 83, no. 1, pp24–45

33. Johnson, Lyndon. (1964) "Remarks at the ground-breaking ceremony for the Florida Cross-State Barge Canal, February 27, 1964," The American Presidency Project, www. presidency.ucsb.edu/ws/index.php?pid=26085, accessed 21 March 2010

34. St. Johns River Water Management District. (2009) *Water Supply Assessment 2008* [draft], St. Johns River Water Management District, Palatka, FL, p33

35. South Florida, Southwest Florida, and St. Johns River Water Management Districts. (2008) *Recommended Action Plan for the Central Florida Coordination Area*, St. Johns River Water Management District, Palatka, FL

36. Love, G. Warren, and Clyde S. Conover. (1986) *Water for Florida Cities*, U.S. Geological Survey, Water-Resources Investigation Report 86-4122, U.S. Geological Survey, Washington, DC, p23

37. St. Johns River Water Management District. (2008) *Executive Summary of the Development of Minimum Flows and Levels on the St. Johns River and Evaluation of the Effects of Potential Water Withdrawals*, St. Johns River Water Management District, Palatka, FL, pp1–14

38. *Johnston, J. Lawrence, St. Johns Riverkeeper, Inc., and St. Johns County v. St. Johns River Water Management District*, Case Nos. 08-1316, 08-1317, and 08-1318, January 12, 2009

39. Patterson, Steve. (2009) Quoting Susan Hughes in "St. Johns water withdrawal permit approved," *Florida Times Union*, 14 April

40. National Research Council, Committee to Review the St. Johns River Water Supply Impact Study. (2009) *Report 1*, National Research Council, Washington, DC, p3

CHAPTER NINE

South Florida Water Management District: Managing Nature?

T he South Florida Water Management District is nothing if not ambitious. Its vision is to be the "world's premier water resource agency."[1] The district may need to be the world's best to manage, protect, and enhance the region's unique but badly damaged water resources. Operating under the same 1972 Water Resources Act as the other four water management districts, the South Florida Water Management District has evolved into a substantially different institution. The district has 31 percent of the state's land area, 41 percent of its people (7.5 million), and 52 percent of the water withdrawn statewide. Its budget usually equals that of the other four water management districts combined. It owns or controls 1.4 million acres in the region. The water resources in south Florida are different from the rest of the state and the problems there are uniquely challenging.

Of the five water management districts, South Florida has the most extensive water resource damage, the greatest conflicts between water users, and the most complicated problems. Those problems are the heritage of more than a century of determined human effort to convert south Florida natural water bodies into forms more suitable for economic exploitation. The highly structural approach to water management in south Florida is, in part, a response to particular hydrological features. Rainfall averages 53 inches a year, including occasional storms that release enough rain to flood large regions. Average rainfall is large but highly seasonal. During the three winter months of December, January, and February, the district receives on average only 12 percent of its annual rainfall. People in the area sometimes say that there are only two seasons of the year in south Florida: wet and dry.

Topographic relief is extraordinarily limited. Only a small northern section rises more than 100 feet above sea level, while most of the district from Lake Okeechobee south is less than 20 feet above the ocean. Much of it is less than 10 feet above sea level. The low relief creates a water system that swings irregularly from flood to drought. Low relief also makes the region uniquely vulnerable to the threat of sea level rise induced by long-term climate change.

The internationally recognized Everglades is only the downstream section of an enormous 220-mile-long watershed beginning on the south side of metropolitan Orlando in central Florida. Much of this hydrologic system remains, even after many decades of major alterations. Far upstream, water begins moving southward near what is now the tourist district on International Drive in Orlando. These waters eventually feed into Lake Kissimmee and then into the Kissimmee River. On its way southward, the water that once ran freely from the Kissimmee River through Lake Okeechobee and the Everglades to Florida Bay must now trace its way down an interconnected system of obstructing dikes, levees, canals, and pump stations. The biggest stop along the way is Lake Okeechobee, sometimes called "Florida's Liquid Heart."

Lake Okeechobee is the largest lake in Florida. About 6,000 years old, its 730 square miles support a nationally famous sport fishery and many threatened and endangered species. Although large in surface area, the lake is very shallow. At its deepest part, its bottom is no higher than sea level. The maximum recorded elevation of the water in the lake is only 18.77 feet above sea level, set in 1947. That level would not be allowed to occur today due to danger of collapse of the encircling Herbert Hoover Dike. If the lake threatened to reach the 1947 level, water managers would open the flood gates and rush water out to the east and west (despite the damaging results to the St. Lucie and Caloosahatchee River estuaries). As they say in the region, much fresh water is "wasted to tide." Many decades of drainage and flood control activities work well in removing water from the southern end of the peninsula much more quickly than natural runoff.

A visitor to Lake Okeechobee can see the lake only after climbing to the top of the dike surrounding it. The dike created an artificial system that allows water managers to treat the lake as a water supply reservoir rather than a natural lake. Within the managed lake are serious water quality and natural system problems. Natural concentrations of phosphorus in Lake Okeechobee were in the order of 40 parts per billion, but human activities have driven it to over 150 ppb.[2] It is hard to see how any system of water quality management in the Everglades can work unless the phosphorus in Okeechobee is cleaned up. The water quality problems of Lake Okeechobee are passed on to the constructed Everglades Agricultural Area on the south shore, which itself exports pollutants southward to constructed Water Conservation Areas and then to Everglades National Park.

Today, only a remnant of the "River of Grass" is preserved within Everglades National Park. Established in 1947, it is the third largest national park in the lower 48 states and the largest wilderness area in the eastern United States. Its 1.5 million acres have been designated as an International Biosphere, a World Heritage Site, and an Outstanding Florida Water. In 2009 it was named as one of the 25 national parks most vulnerable to climate change.[3]

At the southern end of the Florida peninsula, the sheetflow out of the Everglades feeds Florida Bay, a shallow 770-square-mile estuary of mangrove islands and open water. Like the Everglades, North America's largest subtropical estuary is under biological stress. For decades, this giant estuary has been starved of water in the right amounts and at the right times, which has harmed seagrass beds, the nursery functions for many fisheries, and the sport fishing industry itself. The southeast side of Florida Bay is marked by the archipelago of the Florida Keys, a series of limestone islands that barely reach above sea level but which receive 3 million visitors a year and are home to more than 70,000 people. Drinking water for Keys residents is delivered by the Florida Keys Aqueduct Authority. Under a permit from the South Florida Water Management District, the Authority pumps drinking water from a wellfield on the Florida mainland next to Everglades National Park through a 130-mile pipeline to Key West.

From near Disney World in Orlando to Lake Okeechobee to Florida Bay to Key West, the classic Florida growth management pattern prevails: Floridians have diked, drained, pumped, leveed, dammed, flooded, and otherwise tried to reshape south Florida water resources. The South Florida District, consequently, operates what it terms "one of the world's largest public works projects."[4] That regional set of structures includes more than 2,600 miles of canals and levees, 532 water control structures, and 63 pump stations. Of the state's five water management districts, only South Florida ever expresses water volume in "acre-feet" because only that district thinks in terms of how much floodwater its pumping system can remove per day from flooded cities and farms.

The pump stations move water with a combined force of more than 100,000 horsepower. For example, the water management district operates the giant S-5A pump station it built with the U.S. Army Corps of Engineers in 1955. Diesel engines originally placed in World War II-era submarines power the six main pumps of S-5A. This single pump station can move a total of 3.1 billion gallons of water a day.[5] But S-5A is only one of the enormous pump stations used to adapt the region to human uses. Most of the stations remove flood water from agricultural and urban lands. In recent years, additional pump stations were added to help operate the constructed Stormwater Treatment Areas (STAs) as part of Everglades restoration efforts. The STAs built so far have more than a dozen pump stations, more than 200 water control structures, and hundreds of miles of levees and canals.[6]

The district pumps move gargantuan amounts of water. During flood periods, district pump stations designated as S-5A, S-6, G-310, G-335, S-319, and S-362 together can move 12.3 billion gallons a day of water. That is 78 percent more than the average daily flow of the Suwannee River's 6.9 billion gallons a day. Although the peak pumping capacity is astounding, even the average amount of water moved by the district's pump stations is immense. Over the last decade, they pumped an annual average of 2.61 billion gallons of water a day. Numbers like these are hard to place at a human scale. Another way of expressing the same volume of water is more than 350 gallons of water a day for every resident in the South Florida Water Management District. That is very nearly twice the amount of water used daily in the district for public water supply.[7]

The extensive physical alteration of the Everglades region has inflicted major environmental damages. Perhaps the most visible harm has been to the region's famous populations of wading birds. Some early accounts estimated numbers at more than 2 million – the largest numbers in North America and perhaps even the world.[8] Around 1900, hunters devastated the populations of snowy egrets, roseate spoonbills, and reddish egrets for the millinery trade. The hat fashion finally passed, but what did not change in the next decades was the Florida passion to drain the wetlands on which the birds depended. Unnatural patterns of water flow deprived the birds of the aquatic prey they needed for themselves and their nestlings. Bird populations decreased 90 percent from their heights of the late nineteenth century.

It is not only the physical drainage and flood control works that transformed the ecosystem, however. The natural assemblage of species in south Florida is also under biological siege, more so here than in any other part of Florida. Irresponsible owners frequently release tropical animal pets into the south Florida environment. About one-quarter of the fish in Florida are nonnative – a higher percentage than in any other state. There are more introduced animal species in Florida than in any other part of the United States. The new animals frequently do not encounter the biological controls that limited their numbers in their native environments. South Florida is especially affected. For example, about 99,000 Burmese pythons were legally imported into the United States between 1996 and 2006.[9] Thousands more are bred in captivity each year. The pythons can grow to be more than 20 feet long and weigh more than 200 pounds. They eat native mammals, fish, birds, reptiles, and amphibians. The subtropical environment of south Florida is proving to be a perfect home for these excellent swimmers. More than 1,200 Burmese pythons were removed from Everglades National Park between 2000 and 2009. The South Florida Water Management District has assigned one worker to capture or kill Burmese pythons. Since 2004, Bob Hill has handled more than 300 pythons that he finds, in part, by picking up whiffs of the musk given off by the snakes. He once removed an 11-footer from a pump station bathroom. Hill is directed to carry a Winchester shotgun with No. 4 steel shot and usually aims for the snake's head.[10] According to a 2009 U.S. Geological Survey study, the Burmese python and other giant constrictor snakes introduced into south Florida pose "an exceptional threat to the integrity of native ecosystems."[11] Fortunately, the unusually cold winters of 2009 and 2010 appear to have reduced population levels somewhat but the Burmese pythons and other tropical invaders are expected to rebound.

Megafauna like Burmese pythons get media attention, but it is not just large snakes that impair south Florida aquatic ecosystems. The survival of the snail kite is threatened because the small native apple snail on which it feeds is being supplanted by much larger varieties of South American apple snails. The introduced species of snail are much larger and much harder for the kite to consume.

Exotic and invasive plants may be an even bigger and longer-term water resource problem in south Florida than introduced animals. In subtropical south Florida, invasive and introduced plant species are more numerous and cause greater problems than in any other part of the state. The South Florida Water Management District spends almost $20 million a year on invasive species management.[12] For example, the Melaleuca tree was intentionally brought from Australia at the beginning of the twentieth century in an effort to dry up wetlands. It grows in vast monocultural stands that choke out native vegetation. Melaleuca has very limited wildlife value. The South Florida Water Management District spent more than $1 million dollars in 2008 herbiciding or otherwise destroying Melaleucas on more than 150,000 acres of their land. The good news is that the district, working in cooperation with other agencies, has made dramatic progress in recent years in reducing the Melaleuca infestation in the Water Conservation Areas and along the fringes of Lake Okeechobee.

Another recent dangerous plant invader is Old World climbing fern, first seen in 1965. Native to southern Africa and southeast Asia, this plant grows rapidly and densely on the land surface but its vines also grow to the tops of tall trees. Unlike some other invasive species, Lygodium can take over natural ecosystems not much disturbed by human alterations. For example, it has covered natural tree islands in the Water Conservation Areas. During natural or prescribed fires, the dense vines of Lygodium act as "fire ladders" and can destroy trees that otherwise are adapted to survive fires. According to the South Florida Water Management District, this single species "could potentially overtake most of the southern peninsula of Florida."[13]

Despite all of the district's control efforts, there is little hope of exterminating invasive plants in the southern part of the Florida peninsula. The management goal is "maintenance control," which means keeping invasives from causing significant damage. That effort means permanent ongoing expenditures to prevent population expansions. In a striking historical reversal, the water management district with the longest history of converting natural ecosystems into artificial ones is the same district spending the most to protect natural systems from the onslaught of biological invaders.

FIRST ATTEMPTS TO GET RID OF THE WATER: BEGINNINGS TO 1928

From the beginning, state government tried to promote canals and drainage. Even before admission to the Union, Florida's territorial delegate to Congress offered a resolution on December 30, 1842 on the subject:

> Resolved, that the Secretary of War be directed to place before this House such information as can be obtained in relation to the practicability and probable expense of draining the everglades of Florida.[14]

The federal government actions of greatest consequence for Florida water resources in the nineteenth century proved not to be drainage projects but the transfer of millions of acres of national lands to the state. As described in Chapter 2, Origins, the Board of Trustees of the Internal Improvement Trust Fund conveyed million of acres of state land north of Lake Okeechobee to Hamilton Disston on his promise that he would build canals and drain massive amounts of wetlands. He dug more drainage canals than any promoters before him, including a ditch from Lake Okeechobee west to the Gulf of Mexico, which straightened out the Caloosahatchee River. Florida boosters, including Governor Broward, got newspapers to publish claims that the drainage project was the "greatest reclamation

Figure 9.1 *Many rivers, lakes, and wetlands in Florida were dredged and filled in the 19th and 20th centuries; current threats to water resources are more diverse, ranging from water pollution to climate change*

project in the history of the world."[15] The state of Florida, in concert with regional drainage districts, built 6 major drainage canals and 50 miles of levees.

By the mid-1920s, Lake Okeechobee had a crude dike on its southern shore that allowed a fringe of agriculture to spring up there. Catastrophe was looming, however, for the agricultural workers by the lake. In 1926 and 1928, tropical storms forced lake water to overtop the primitive dikes on the south side of Lake Okeechobee. More than 2,000 people drowned in the 1928 event. The flood disaster marked out a prominent place in the water management memory of south Florida. As late as 1978, a member of the governing board of the South Florida Water Management District remembered losing most of his family in the 1928 storm. He survived only by clinging to a log drifting through the floodwaters.[16]

BRINGING IN THE FEDERAL GOVERNMENT: 1928–1948

The lesson the state drew from the Okeechobee disaster was not that massive drainage and flood control projects should be rethought. Instead, Florida sought to expand on the existing projects – but at national expense. The Everglades Drainage District simply did not have the resources to take on a project as big as

Lake Okeechobee (and even defaulted on its bonds in 1932). In 1928, presidential candidate Herbert Hoover visited the site of the Okeechobee flood tragedy and committed to help Florida with the Okeechobee project if he were elected. Florida became one of only four southern states to vote for Hoover, and the new president moved to change long-standing federal policy prohibiting national funding of this kind of local project.[17] Congress directed the Army Corps of Engineers to build a stronger dike around Lake Okeechobee. Florida had only to provide the necessary land and 2.5 percent of the $19.1 million cost of construction. In the 1930s, the Corps not only built the Herbert Hoover dike but also enlarged the existing canals to the west (running through the Caloosahatchee River to the Gulf of Mexico) and to the east (running down the St. Lucie River to the Atlantic Ocean). Florida was delighted to see the federal government pay more than 95 percent of the $20 million cost of strengthening the dike.[18]

By dredging out natural riverbeds such as the Miami River and rushing water to the ocean, natural water tables were lowered. Floridians took advantage of the drier land for both agricultural and urban purposes. The lowering of the water table in a region just above sea level had an unintended consequence, however: the intrusion of salty ocean water. No longer held back by freshwater hydrologic pressure, the Atlantic Ocean seeped into the lowered freshwater aquifer by the coast. By 1938, more than a thousand wells near Miami had been "salted up" and forced to relocate further inland. By 1946, saltwater had moved more than 2 miles inland.[19] (Rising sea levels in the twenty-first century may make this problem more severe.)

The next big push to remake south Florida's water map came just after World War II, in response to recurrent bouts of flooding. The Army Corps of Engineers was ready for the next phase of flood control appropriations. On June 8, 1947, District Engineer Col. Willis Teale held a public hearing about possible flood control projects for Belle Glade on the southern side of Lake Okeechobee. Two summers of heavy rains made the plans timely. Much more rain fell after the hearing, making it look almost as if the Corps could control the weather to make their point. As if that were not enough, a major hurricane came ashore on September 17, terrifying the residents around Lake Okeechobee.

Fortunately, the dike held. So did the rain, as two more tropical storms dumped large amounts of rainfall over the entire region in the next month. Miami received 102 inches of rain in 1947. The Caloosahatchee River went over its banks; so did the Kissimmee River and many other waterways. More than 2,000 square miles of south Florida were under water. The Corps was soon ready with even bigger flood control plans, as were Florida politicians. Floridians heavily promoted the need for federal assistance, including a news media campaign. One of their famous productions was a Tentative Report of Flood Damage that featured a drawing of a cow in flood water up to its shoulders, with large tears streaming from its eyes.[20]

Less than a year later, the Corps formally recommended what they called the "Central and Southern Florida Flood Control Project (C&SF Project) to prevent a future flood catastrophe. The C&SF Project was projected to cost $208 million and be a comprehensive solution to the region's flood vulnerability. There would be new canals plus enlargement of existing canals, water storage areas, and a large number of other structures from Orlando south to the Everglades, and from the west coast to the east coast of Florida. The Corps proposed to move more dirt for this effort than in any of their projects since the Panama Canal.[21]

The Corps advised Congress that if "adequate appropriations" occurred, the "entire comprehensive development could be completed under an orderly and efficient construction program in 10 years."[22] Best of all for Florida, state and local interests would have to provide only 18.5 percent of the cost of building the project. Local, state, and national politicians were eager to get Congress to act. U.S. Senator Claude Pepper pointed to how the new Project would, once and for all, address the water control problems of the entire region:

> The time has come when we have got to deal with the flood situation in the Peninsula of Florida as a whole. It is all fundamentally one problem and has got to be approached as a single problem with a single comprehensive program.[23]

Congress authorized the C&SF Project with amazing speed in 1948. The state met its share of the bargain by sweeping away the previous drainage districts and setting up a new Central and Southern Florida Flood Control District (FCD) to act as local sponsor for the federal project. How was the new Central and Southern Florida FCD to be financed? Lamar Johnson, the chief engineer of the Everglades Drainage District, which was slated to be incorporated into the new district, thought that the Everglades Agricultural Area would "unquestionably" receive the greatest benefits of the new flood control project. He and other proponents of the new district favored a tax system in which the taxes paid were proportional to the benefits received. The predecessors to the FCD, like the Everglades Drainage District, had assessed taxes according to the benefit that different parcels of land received from drainage and flood control.[24]

Instead, the legislature imposed a uniform ad valorem tax in which all landowners throughout the district paid in proportion to property value, regardless of whether the project provided them any benefits. That was because the legislators representing the rural agricultural counties, later called "Pork Choppers," refused to create the new district unless the uniform tax system was chosen. That lowered the tax burden for agricultural areas and transferred the cost of the new FCD to urban areas.[25]

State government also changed its longstanding policy for the FCD. The legislature decided to contribute general revenue funds for building the C&SF Project, which had never been done before. Luther Carter, in his 1974 history of land and water policy in Florida, concluded that the cost sharing for the FCD project favored agricultural interests, especially large corporate farms of the Everglades Agricultural Area, "in an outrageously unfair way." Metropolitan areas on the lower east coast paid a majority of the property taxes of building the project even though they received little flood control benefit from it.[26] Six decades later, that same system of "uniform" taxation and substantial state assistance is still used by the South Florida Water Management District.

HIGH TIDE OF THE STRUCTURAL APPROACH: 1948–1971

The new FCD began business as a five-member board on July 14, 1949. After so many previous failures, the *St. Petersburg Times* looked forward in 1950 to a prompt victory in the war against nature:

> For the past century, engineers and scientists from all over the nation have been jousting with nature in a continuous battle for the Everglades. And with the passage of time, they have combined the resources of many departments of government and private enterprise to a point where nature is about to take a licking.[27]

The tactical plan for the assault was written in the federal law authorizing construction of the C&SF Flood Control Project. The Corps and the FCD moved rapidly. By 1954, they acquired real estate title or flowage easements on 1,354 square miles in Palm Beach, Broward, and Miami-Dade counties for the new Water Conservation Areas in the project. This series of acquisitions preserved, to a degree, almost half of the original Everglades. The rapid progress in the 5 years after 1949 allowed the FCD to anticipate "early completion of the comprehensive plan."[28] In 1955, however, the FCD was forced to admit that the date of completion of the project would not occur until 1965. To meet even this new schedule, the FCD and Corps would have to construct the project at a "much faster" pace.[29] The FCD blamed the federal government for the delay. The federal partner was not matching the state's expenditures, as had been promised in 1949:

> A collateral difficulty has been the reluctance of Congress to authorize features of the [C&SF] project as fast as local and state interests were willing to meet their share of the cost. The steady reduction of the actual federal appropriations for the authorized features in fiscal years 1952, 1953, and 1954 has had a disastrous effect on state planning[.....]The prospect for the next two years appears equally dismal[....]The other problem which confronts the people of the area is whether it may be advantageous to tighten their belts and go forward with certain expenses of the project without waiting for federal cooperation to come up with the level originally proposed by Congress? It is certain that the federal share of the project is lagging and that prospects for accelerated federal attention are dim.[30]

This complaint about the unreliability of federal appropriations is a recurring theme in south Florida water management. By the end of the 1950s, however, the project had built the 700,000-acre Everglades Agricultural Area (EAA) on the south side of Lake Okeechobee. This massive new zone (larger than the state of Rhode Island) was perfect for growing sugarcane and winter vegetables. When farming spread across the EAA, however, so did the magnitude of soil subsidence. In much of the rich, peaty farmland, the soil today is several feet lower than when opened to agriculture and it disappears at the rate of about an inch a year.[31] Some of the organic matter has volatized into the air and the soil level also dropped by simple compaction of the low-density soil. In effect, the very act of farming the organic soils is causing them to vanish. Eventually, farming in the EAA will have to be abandoned.

The Corps and the FCD also built the giant water conservation areas by 1962. They extended and strengthened Herbert Hoover Dike around Lake Okeechobee, built a number of massive pump stations, once again enlarged the channelization of the Caloosahatchee River, and built additional spillways, dams, canals, and navigation locks. The region's original surface hydrology, in which water followed the laws of gravity, was being transformed into a human-dominated system that could force water to flow in almost any direction the water managers desired.

The Corps published a series of color-coded maps showing in green the project works not yet built and in red the components of the C&SF Project already constructed. Over time, red lines became more widespread over the map of south Florida. The expected 1965 date of completion was not met, however. In 1966, the FCD forecast that the project was 5 or 6 years from completion, at a cost of close to $500 million.[32] (That deadline, too, was missed. Four decades later, Florida is still billions of dollars, and decades, away from again fixing the 1948 project.)

The largest C&SF Project component remaining in 1962 was turning the beautiful Kissimmee River into the unlovely Canal C-38. Despite opposition from the U.S. Fish and Wildlife Service and the Florida Game and Fresh Water Fish Commission, the FCD and the Army Corps of Engineers rapidly channelized the Kissimmee. From 1962 to 1971 they converted the many meanders of the 103-mile-long Kissimmee River into a ruler-straight canal 56 miles long, 30 feet deep, and 300 feet wide. Canal C-38 was a plumber's masterpiece and an environmental disaster. The canal superhighway damaged about two-thirds of the riverine wetlands, depleted a popular recreational fishery, and reduced migratory waterfowl by 92 percent. After channelization, the old river channel still wended back and forth across the old floodplain, but in a combination of dry land and stagnant pools cut off from the life of the natural river. Agricultural interests quickly moved into the floodplain.

CHANGING COURSE: 1971–1999

As the FCD entered the 1970s, it appeared to be as powerful and stable as other contemporary water agencies and institutions. Like the dam engineers and civic boosters of the western United States whom Marc Reisner described in *Cadillac Desert* (1993; Penguin, New York), the FCD was physically remaking south Florida. As in the western experience, the FCD depended heavily upon the federal government for financial support. By 1972, for example, it had received more than three times as many dollars from the federal government as from the state.[33] Like the Bureau of Reclamation in the U.S. Department of Interior, the FCD and the Corps of Engineers approached water management primarily as a technical and engineering problem. Indeed, the managing official for the FCD was called the "chief engineer" into the 1970s. Water was often treated as a problem to be attacked rather than a feature of the natural and human world to be always preserved and treasured.

Donald Worster, a leading historian of the American West and its water management, called that region "the greatest hydraulic society ever built in history."[34] South Florida is far from arid, but the C&SF Flood Control District had almost as much influence over its region as the California Water Project and the U.S. Bureau of Reclamation had over theirs. However, Floridians were beginning to demand more from water managers than engineered domination over their extraordinary water resources.

In 1971, the same year that *Time* magazine called "the environment" the "Issue of the Year," the Corps and the FCD completed the channelization of the Kissimmee. Everyone understood that converting the river into a canal would cause major losses of wetlands, fish, and wildlife, but those costs now seemed unacceptable. A significant drought in south Florida also focused attention on what would decades later be called water resource sustainability. The 1971 Governor's Conference on Water Management in South Florida (described in Chapter 2, Origins), was a prominent milestone of change. The conferees recommended to Governor Askew that there should be "no further draining of wetlands for any purpose."[35] Some of their recommendations went far beyond the scope of what the legislature was willing to enact the next year. The 1972 Florida legislature did much, however. In its own "Year of the Environment," it passed four major environmental laws, including the Water Resources Act of 1972. The new South Florida Water Management district hired Professor Frank Maloney, the primary author of the Model Water Code, to help them draft the water use permitting rules necessary to implement the Act.[36]

Even with changing attitudes toward water management (and new laws), it was not clear whether there was any prospect of undoing the channelization of the Kissimmee River. Agriculture was the primary beneficiary of channelization and continued its dominance in south Florida water management. So many of the board members of the new South Florida Water Management District were oriented toward agricultural interests that Robert Clark, the governing board chair in the 1970s, said it didn't matter much for agriculture if the political orientation of board members was Democratic or Republican.[37] "Big Ag" usually got what they wanted. Nonetheless, some began to dream of restoring the Kissimmee.

Restoration of the Kissimmee River would be the most ambitious and costly such project ever attempted in the world. It was remarkable, therefore, that the state's Coordinating Council on Kissimmee Restoration formally voted in 1983 for "partial backfilling." However, it was not until 1992 that Congress authorized Kissimmee River restoration and not until 1999 that backfilling actually began putting the braids back in the river's course. Only in 2006 was the last acre acquired for restoration. The project is not expected to be completed until at least 2013. More than 39 square miles of river floodplain will be restored or substantially improved, at a cost approaching $600 million – many times more than that expended in the original channelization.[38]

Supporting Kissimmee River and Everglades restoration became a bipartisan political winner in Florida for both governors and presidential candidates. Democratic Governor Reuben Askew strongly promoted both the creation of water management districts in 1972 and their indispensable taxing source in the

1976 constitutional referendum. Governor Bob Graham, Askew's successor and also a Democrat, announced his Save Our Everglades program in 1983, calling for the Everglades to look by 2000 more like it did in 1900. Both governors won reelection. Estus Whitfield (environmental adviser to Bob Graham and two subsequent governors) explained the origins of the Save Our Everglades Program and the Everglades Coalition:

> What led up to [the] Save Our Everglades program was a major conservation effort by the Florida Wildlife Federation led by Johnny Jones, influenced by Art Marshall. There was Charles Lee. Charles Lee was young at the time, [and] was with the [Florida] Audubon Society. Then there was the Sierra Club at the time. They were the main organization, along with the Audubon and the Wildlife Federation. They touched a lot of the other grassroots organizations. When Graham kicked off the Save Our Everglades program in 1983, it was with their blessing. They were right behind us because it was to restore the Kissimmee, rectify the drainage problems in the Everglades, restore Everglades National Park, a lot of good features, and they loved it. Graham went to Washington in 1983 or 1984 and met with all the national conservation organizations, all the major conservation organizations. I went with him and took big maps. We made presentations on [the] Save [Our] Everglades Program. The Florida organizations came to that presentation, four or five of them. Charles Lee was instrumental in that. It was the National Parks Conservation Association, the National Sierra, the National Wildlife Federation. About fifteen of those organizations. They decided on that day while we sat there in the National Parks Conservation Association conference room in Washington D.C., to form the Everglades Coalition.[39]

Since the initial annual meeting of the Everglades Coalition in January 1985, each Florida Governor has made it a point to attend the first meeting that occurs after inauguration as Governor. The record of attendance was broken by new Governor Rick Scott in 2011.

With strong political support, Kissimmee restoration received many millions of dollars in state and federal appropriations. The Army Corps also began to recognize the changing world of water management in Florida. As Everglades restoration proposals came to the fore, Corps officials recognized that the Kissimmee restoration project was a kind of template for Everglades restoration. It was not just the Kissimmee River system and the Everglades that needed attention, however. Other problems were intensifying in Lake Okeechobee and ecosystems to the east, west, and south of the big lake. Under the presidential administration of George H. W. Bush, the acting U.S. attorney in Miami sued the South Florida Water Management District and the Florida Department of Environmental Regulation (now the Department of Environmental Protection) in 1988 for violating state water quality standards in Everglades National Park and Loxahatchee National Wildlife Refuge. Dexter Lehtinen's lawsuit, filed without checking with his superiors in the Justice Department, became an epic legal battle.

Governor Lawton Chiles, a Democrat and former U.S. senator, tried to end the litigation in 1991. He appeared in the judge's courtroom to surrender:

I came here today convinced that continuing the litigation does little to solve the problems or restore the Everglades. I am more convinced than ever of that … We talked about water in the glass … I am ready to stipulate today that water is dirty. I think that is [what this is] about, Your Honor, is how do we get clean water? What is the fastest way to do that? I am here and I brought my sword. I want to find out who I can give that sword to and I want to be able to give that sword up and have our troops start the reparation, the clean up [….] We want to surrender. We want to plead that the water is dirty.[40]

That "surrender" helped move the parties toward a settlement agreement but it was two more years before the litigants signed a Statement of Principles to suspend the lawsuits and work together on a detailed technical plan for restoration. Congress also played a major role in the evolution of ideas about the Everglades. It had already authorized in 1992 a "restudy" of the C&SF Flood Control Project. Little new was expected from the Corps in conducting the restudy because they had concluded in 1989 that "no federal action is advisable at this time" in rethinking the C&SF Project.[41] To increase the likelihood of federal action, Governor Chiles set up the Governer's Commission for a Sustainable South Florida in 1994. He persuaded Richard Pettigrew to chair the 51-member commission. Pettigrew had been speaker of the Florida House in the early 1970s and oversaw the creation of the modern Water Resources Act in 1972. According to Pettigrew, his service then occurred during a "brief period of several years, almost a decade, in which the lobbyist influence was dramatically reduced. Now, it has come back, and now the lobbyists are just running the place."[42] Nonetheless, he was willing to try to bring together the diverse stakeholders of south Florida and help design Everglades restoration.

Pettigrew was able to report to Governor Chiles on October 1, 1995 that the commission had achieved unanimous agreement on the fundamental direction the restudy should take. Governor Chiles rewarded the first report of the commission with an additional assignment: reaching consensus on an even more detailed conceptual plan for Everglades restoration and sustainability in south Florida. In August 1996, the Governor's Commission succeeded and, again by consensus, approved a general guide to modifying the project. The commission grouped 40 "preferred options" into 13 "thematic concepts" to fix south Florida water resource problems. Their ideas fed directly into the restudy of the C&SF Florida Flood Control Project authorized by Congress and conducted by the Army Corps of Engineers.[43]

Under intense pressure to complete a feasibility study, the Corps and the water management district met the schedule with a 10-volume opus, weighing down bookshelves with 23 pounds of paper. The Corps' Comprehensive Everglades Restoration Plan ("CERP") had 68 separate project components aimed at "getting the water right" in regard to quantity, timing, quality, and distribution.[44] The natural system would get more of the clean water it needed (and when it needed it), farms and cities would get more water supply for the future, and flood protection would be maintained. The restudy estimated that 80 percent of the "new" water would go to the natural environment. It would be the largest environmental restoration project ever attempted on the planet.

Implementation of CERP would require decades. If Congress approved the new Corps plan for south Florida, new surface water reservoirs would occupy more than 180,000 acres of land in the region. That would allow enormous treatment marshes to be built and also enable Lake Okeechobee to cycle less frequently between damaging extremes of drought and flood. It would cost at least $7.8 billion, not including Kissimmee River dechannelization and other important projects in south Florida. The federal government agreed to pay 50 percent of the costs of building CERP; the state and the South Florida Water Management District would have to pick up the other half.

The Corps firmly recommended the new approach to Congress in 1999. The chief of the Army Corps of Engineers emphasized that "The primary and overarching purpose of the Comprehensive Plan is to restore the south Florida ecosystem." The Assistant Secretary of the Army for Civil Works (the civilian overseer of the Corps) called the Everglades an "American treasure on par with the Grand Canyon, Yellowstone, and California's ancient redwoods." Vice President Al Gore began discussing the Everglades legislation with President Bill Clinton at their weekly lunch meetings.[45]

Congress proved ready to endorse Everglades restoration. The diverse stakeholders who helped write the Everglades restoration plan mostly held together and lobbied cooperatively for its inclusion in the Water Resources Development Act of 2000. However, some environmental groups faulted the plan as ceding too much to the demands of farmers and cities for more water supply and more flood protection. Friends of the Everglades, an organization founded by Marjory Stoneman Douglas, opposed passage of CERP.[46] Despite some opposition, Congress included CERP in the Water Resources Development Act of 2000 and laid down as national policy the main purposes of the program:

> The overarching objective of the Plan is the restoration, preservation, and protection of the South Florida ecosystem while providing for other water-related needs of the system, including water supply and flood protection.[47]

THE FIRST DECADE OF CERP: 1999–2009

Early success with Kissimmee River restoration gave some confidence that the Everglades, too, could be restored. However, Everglades restoration was expected to be much more difficult than the Kissimmee. The river was a much smaller project than the Everglades system, had suffered its worst damage only recently, and was much simpler. Tommy Strowd, operations director of the South Florida Water Management District, explained, "The Kissimmee restoration is like 'Oops, we dropped something, let's pick it up.' It's immediate, but the Everglades is different. We can't just go back to nature."[48]

Just like in the 1950s, the federal government did not keep financial pace with the state of Florida and the South Florida Water Management District. Congress failed to appropriate the dollars to match the local sponsor's contributions. Within a few years of the enactment of CERP, environmentalists, cities, the Congressional General Accounting Office, the National Research Council, and others all criticized the slow pace of implementing CERP. The delays and criticism caused serious doubt about whether CERP would ever get on track. In 2007, for example, the General Accounting Office found that a "core group of projects that are critical to the overall success of the [Everglades restoration] effort are behind schedule or not yet started. Florida had spent $4.8 billion between 1999 and 2006 on Everglades projects, the federal government had contributed only $2.3 billion."[49] Complaints in Florida about the slow pace of federal appropriations for CERP echoed complaints in the 1950s that the feds were not living up to their end of the bargain on the original C&SF Flood Control Project.

Even with funding slowdowns, the water management district and Corps (with state assistance) constructed or initiated a number of major CERP and other restoration projects while Republican governor Jeb Bush was in office from 1999 to 2007. For example, the district built 64 square miles of treatment marsh near Lake Okeechobee to improve the quality of the water headed toward Everglades National Park and coastal estuaries. These massive Stormwater Treatment Areas (STAs) capture hundreds of thousands of acre-feet of water and run it through a treatment system of algae and aquatic plants. The STAs are intended to remove 80 percent of the phosphorus in the water before releasing it downstream for other purposes. Stormwater Treatment Area 3/4 is said to be the largest treatment marsh in the world, with 29 miles of canals and 31 miles of levees. At 26 square miles, it has a land footprint larger than the island of Manhattan in New York City.

Despite major delays, CERP continued to score some successes. In 2007, Congress passed a Water Resources Development Act for the first time since 2000 when it had authorized the Comprehensive Everglades Restoration Plan. The bill included specific authorization to build some $2 billion worth of the first CERP projects.

THE BIG SUGAR DEAL

The Plan enacted by Congress made substantial progress in the first decade, although it was moving more slowly than hoped. Then, like previous disruptive events in south Florida, an opportunity presented itself. In 2008 U.S. Sugar Corporation dispatched two lobbyists to meet with Governor Crist to complain about the South Florida Water Management District.[50] One of their gripes was that the district's governing board, led by Crist's new appointees, had reversed a decades-old policy of allowing farms in the Everglades Agricultural Area to backpump their excess polluted farm water into Lake Okeechobee. The governor was being shown maps of pump stations when he asked a surprising question: Why not just buy out the entire company?[51]

Company executives claimed they were "stunned" by the initial proposal to buy them out. However, they soon proved to be receptive to a deal – if the price was right. After a series of negotiations, the district proposed to buy, as a package,

292 square miles of U.S. Sugar land, including the company's new sugar mill, a sugar refinery, 20,000 acres of citrus near the Everglades Agricultural Area, a citrus processing plant, and 200 miles of railroad. But it would cost an estimated $1.75 billion at a time when a national recession was drastically shrinking both state and district budgets.

If consummated, this "big sugar deal" would reduce loadings from a major source of pollution to the Everglades, but there would be many other benefits too. The massive acquisition also would create new options for CERP. Water could be stored on the acquired land during wet times, which would reduce harmful flood discharges from Lake Okeechobee to coastal estuaries. The stored water would also be available for use by humans and natural systems during dry times. However, many other carefully designed CERP components would become obsolete if the acquisition went through; much of CERP as envisioned when Congress authorized it less than a decade before would have to be reconfigured or abandoned.

In the face of the national recession and declining tax revenues for the water management district and state government, the size of the proposed acquisition soon began to shrink. First, the state and the South Florida Water Management District decided they did not need the sugar mill, the railroad, or other capital assets – they wanted just the 187,000 acres of land. As negotiations continued, the deal shrank again to 73,000 acres and $536 million. The district would still be acquiring parcels of land almost twice the size of Orlando and it would also have the right to acquire the remaining 107,500 acres within 10 years. Even the reduced project was a hard sell for the governing board in fiscally constrained times. Farm interests, some legislators, and other opponents urged that the acquisition be canceled or postponed. Environmentalists urged completing the deal as a bold new step toward Everglades restoration. Long-time Florida environmentalist Nat Reed said that the district simply must acquire the property:

> This is the dream of dreams that Marjory Stoneman Douglas proposed and everybody laughed at because nobody thought it would ever be possible. She'd say buy it. She'd say, 'Boys, twenty years from now nobody will remember the defects and everybody will say my God, they had the courage to save the Everglades.'[52]

On December 16, 2008, the governing board voted 4–3 to accept the "final" offer but added one more condition that could terminate the deal. The district wanted to reserve the right to back out before the final contract was signed if fiscal conditions continued to deteriorate and if it proved unable to perform its core water supply and flood control missions. U.S. Sugar accepted the offer the same day of the board vote, calling the added condition only a "nonmaterial modification" of the contract offer.[53] The governor called a news conference the next day to commend the vote and call on a higher authority than even Marjory Stoneman Douglas:

> There's no doubt in my mind, God is looking down on us and he is happy. It's not cheap, but neither is this earth. And neither are the Everglades. Nor should they be."[54]

The "big sugar deal" was considered before the national real estate bubble burst, which hit Florida especially hard. By early 2010, the district's financial adviser reported that buying the U.S. Sugar land would be possible only by making very difficult decisions on priorities, in light of declining property tax collections.[55] As ad valorem tax revenues have declined $150 million a year, state contributions to Everglades restoration have also gone down dramatically. The district refused to consider raising property tax rates. In August, 2010, the South Florida District governing board and U.S. Sugar again shrank the deal, this time to only 26,800 acres (much of it in citrus operations) costing $197 million in cash that the district had in the bank. The district retained an option to buy more land in the future if fiscal conditions improved (or if the federal government came up with the necessary funds). Florida gubernatorial candidate Rick Scott held a press conference outside the district's headquarters on the same day to express his opposition to the proposed acquisitions, calling it an "irresponsible" use of taxpayers' money and a "sweetheart deal" for U.S. Sugar.[56] Scott followed up this campaign statement in his 2011 budget proposal by successfully recommending that state appropriations for Everglades restoration be dramatically cut. He also appointed to the SFWMD Governing Board an organizer of a "Tea Party" that opposed the U.S. Sugar acquisition.

It appears that the original 26,800 acres might be very valuable in improving water quality and complying with orders of federal judges. As part of two decades of Everglades litigation, John M. Barkett (appointed as Special Master for federal judge Federico Moreno) concluded in August 2010 that the U.S. Sugar acquisition had "radically altered the Everglades landscape" for CERP projects.[57] He concluded that the acquisition, in combination with the Everglades Agricultural Area A-1 Reservoir project that the district had partially completed, presents new opportunities for Everglades restoration. The Special Master's finding matched one by the U.S. Environmental Protection Agency only a month later. The EPA required a series of specific measures for the state of Florida and the South Florida Water Management District to meet water quality standards in the Everglades.[58] This would include the construction of an additional 42,000 acres of stormwater treatment areas. The U.S. Sugar acquisition may be a key part of these new projects.

A LONG-TERM VISION FOR THE COMPREHENSIVE EVERGLADES RESTORATION PLAN?

In 2010, the stars seemed to be aligning for CERP. Critically important federal officials in the Obama administration appear to be firmly on the side of Everglades restoration. These include:

- *Carol Browner*, President Obama's energy and climate "czar" and former administrator of the U.S. Environmental Protection Agency. She is a Florida native and the former secretary of the Florida Department of Environmental Protection.

- *Colonel Terrence "Rock" Salt,* the deputy assistant secretary of the Army for Civil Works. In that role, he oversees the entire Army Corps of Engineers, the principal federal agency for implementing CERP. He previously served as the Corps district engineer in Jacksonville and the executive director of the South Florida Ecosystem Restoration Task Force.
- *Colonel Alfred A. Pantano, Jr.,* the district engineer for the Corps Everglades' project in mid-2009. Just before assuming those duties, he completed a paper for a master's degree at the Army War College, titled "Everglades: The Catalyst to Combat the World Water Crisis." According to his thesis paper:

> Mankind cannot even afford to wait to complete the CERP within the timeline projected. The political will to act now on the largest restoration plan in the history of the world is as vital to life as the water it will provide. Success of the CERP will result in immediate benefits to Florida. More importantly, it will catalyze the will and ambitions of leaders across the United States and the world to act. Failure to act is not an option if mankind cares to survive.[59]

For CERP, this set of officials was a remarkable "dream team." Moreover, federal officials also finally had big dollars to spend on CERP. In what the *Miami Herald* called a "river of cash," Congress allocated hundreds of millions of dollars for Everglades restoration.[60] Almost $200 million dollars for Everglades projects was in President Obama's "stimulus package" that Congress enacted in 2009.

A decade after Congress approved CERP, restoration appears to be struggling but has seen marked successes. It will require decades, however, to determine whether the restoration is successful and, in light of Florida's current government revenue crisis, may be possible only with expanded federal assistance.

LOOKING BACKWARD, LOOKING FORWARD

Despite great efforts and enormous expenditures, it will be hardest to overcome water management obstacles in the South Florida Water Management District. The natural and human challenges are daunting. Several generations of Floridians did as much, or even more, damage in this region as anywhere else in the state. The water resources of the region were immense, diverse, and relatively easy to drain and channelize. Floridians devoted a century to permanently altering the water map of south Florida. Water management in south Florida operates with a sense of history because much of what is planned for the future is fixing what was done in the past. Changing what is "wrong" to what needs to be "right" is the fundamental goal of CERP. That is to occur under the broad framework of what restoration planners composed as a four-part definition of "getting the water right":[61]

- *Quality. The right water quality.* Excessive amounts of mercury, phosphorus, and other pollutants harm water resources and contaminate the living organisms that depend on them. Improving water quality is a prime purpose of the Stormwater Treatment Areas and indispensable for a healthy environment.

- *Quantity. The right amount of water.* Much less water flows through much of the system today than historically. CERP will store water now rushed to tide and release it at the right time for natural system needs, as well as for agricultural and urban demands. Project components will increase the extent of natural areas and enhance the availability of water supplies. (The last three months of 2010 set a record low for precipitation, which underlines the ongoing importance of improved water management for human and natural uses.[62])
- *Timing. The right timing.* The natural cycle of flood and drought is what created south Florida ecosystems. Those natural hydroperiods were altered in the quest for settlement and agricultural development. Restoring natural timing, or mimicking it as closely as possible, is key to the survival of the many different ecosystems of south Florida.
- *Distribution. The right place.* Half of the original Everglades has been converted to other land uses and the water that fed it has been diverted to other places. Levees and canals force water in new directions. It will be necessary to remove some of the current canals and levees and undertake other structural works to make the water flow into the places that need ecosystem maintenance and restoration.

These are, at heart, very simple principles. Hydrology created the ecosystems of south Florida. The four elements of quantity, quality, timing, and distribution are good principles, but they are not enough. Plans for CERP always had two more principles:

- *Restore, Preserve, and Protect Natural Habitats and Species.* This principle recognizes that the biological threat to the ecosystems of south Florida depends on more than restoration of the hydrology. Invasive species can damage even a system with restored water quantity, quality, timing and distribution.
- *Foster Compatibility of the Built and Natural Systems.* Long-term ecosystem health appears possible only if the natural parts of the region are managed in concert with the highly developed agricultural, industrial, and urban sections. Water supply for farms and cities is interrelated with water for the environment.

Restoration is a process, not a product, and decades will be required for CERP to reach its goals. In the meantime, the biggest hydrologic threat to South Florida water resources is global climate change. South Florida's extremely low land surface means that only a few feet of sea level rise would inundate many coastal cities and much of the Everglades. A higher sea level is far from the only water problem posed by climate change, however. America's only continental coral reefs off southeast Florida may disappear due to the twin climate change impacts of higher sea temperatures and a more acidic ocean. The hurricanes that batter south Florida cities and ecosystems more often than any other part of the United States may become even more frequent or more severe. The already high number of droughts and floods may become more common and more intense.

In short, what is needed for water resource sustainability is another change in ideas about water. Floridians' concepts of water have changed in the last century and will need to change again. As notions about the values of wetlands have transmuted from noxious wastelands to an emblem of beauty, Florida politicians have paid notice. South Florida water management must mean accepting full responsibility for stewardship over the remaining water resource treasures of south Florida and the entire state.

NOTES

1. South Florida Water Management District. (2010) *Strategic Plan 2009–2019*, p2

2. South Florida Water Management District. (2009) "Lake Okeechobee Protection Program – State of the lake and watershed," in *2009 South Florida Environmental Report*, Chapter 10, p10-2

3. Saunders, Stephen, Tom Easley, and Suzanne Farver. (2009) *National Parks in Peril*, Natural Resources Defense Council (New York) and Rocky Mountain Climate Organization (Denver)

4. South Florida Water Management District. (2006) *Facility and Infrastructure Location Map*

5. South Florida Water Management District. (2010) "Pump stations," https://my.sfwmd.gov/portal/page?_pageid = 1194,2140318,1194_23142802:1194_23144278&_dad = portal&_schema = PORTAL, accessed 14 March 2010

6. South Florida Water Management District. (n.d.) *Stormwater Treatment Areas: Managed Wetlands Improving Everglades Water Quality*

7. South Florida Water Management District. (2009) *Comprehensive Annual Financial Report for Year Ended September 30, 2009*, Table 23, "Water Moved by District Pump Stations," pp vi–27

8. Levin, Ted. (2004) *Liquid Land: A Journey through the Florida Everglades*, University of Georgia Press, Athens, GA, p153

9. Harvey, Rebecca G., Matthew L. Brien, Michael S. Cherkiss, Michael Dorcas, Mike Rochford, Ray W. Snow, and Frank J. Mazzotti. (2009) *Burmese Pythons in South Florida: Scientific Support for Invasive Species Management*, Document WEC242, Wildlife Ecology and Conservation Department, Florida Cooperative Extension Service, Institute of Food and Agricultural Sciences, University of Florida, Gainesville, FL

10. Morgan, Curtis. (2009) "Python hunter: Miami man takes on Glades swamp serpents," *Miami Herald*, 9 August

11. Reed, R. N., and G. H. Rodda. (2009) *Giant Constrictors: Biological and Management Profiles and an Establishment Risk Assessment for Nine Large Species of Pythons, Anacondas, and the Boa Constrictor*, U.S. Geological Survey Open-File Report 2009-1202, U.S. Geological Survey, Washington, DC

12. South Florida Water Management District. (2009) *South Florida Environmental Report: 2009*, Executive Summary, p24

13. South Florida Water Management District. (2009) "The status of nonindigenous species in the south Florida environment," in *South Florida Environmental Report, 2009*, Chapter 9, pp9–41

14. Dovell, Junius Elmore. (1947) "A history of the Everglades of Florida," Ph.D. dissertation, University of North Carolina, Chapel Hill, NC

15. "Greatest reclamation project in the history of the world," *Prescott Journal-Miner*, 24 September 1912

16. Huser, Thomas. (1990) *Into the Fifth Decade: The First Forty Years of the South Florida Water Management District, 1949–1989*, South Florida Water Management District, West Palm Beach, FL, p43

17. Hollander, Gail M. (2008) *Raising Cane in the 'Glades*, University of Chicago Press, Chicago, pp100–101

18. Grunwald, Michael. (2006) *The Swamp*, Simon & Schuster, New York, p199

19. Parker, G., G., Ferguson, S. K. Love, and others. (1955) *Water Resources of Southeastern Florida*, U.S. Geological Survey Water-Supply Paper 1255, U.S. Geological Survey, Washington, DC, p11

20. Florida Everglades Drainage District. (1947) *Tentative Report of Flood Damage*. Picture of cover of report is in Godfrey, Matthew C. and Theodore Catton. (2006) *River of Interests: Water Management in South Florida and the Everglades, 1948–2000*, U.S. Army Corps of Engineers, Washington, DC, p40

21. Warner, Sara. (2005) *Down to the Waterline: Boundaries, Nature, and the Law in Florida*, University of Georgia Press, Athens, GA, p59

22. Huser, op. cit., note 16, p8

23. Quoted in Godfrey, Matthew C. and Theodore Catton. (2006) *River of Interests: Water Management in South Florida and the Everglades, 1948–2000*, Chapter 2, U.S. Army Corps of Engineers, Washington, DC, p33

24. McCally, David. (1999) *The Everglades: An Environmental History*, University Press of Florida, Gainesville, FL, p110

25. Johnson, Lamar. (1973) *Beyond the Fourth Generation*, University Press of Florida, Gainesville, FL, pp161–163

26. Carter, Luther J. (1974) *The Florida Experience: Land and Water Policy in a Growth State*, Resources for the Future, Washington, DC, p96

27. Beaudoin, Mike. (1950) "Nature loses ground in 'Glades, man gains," *St. Petersburg Times*, 12 June

28. Central and Southern Florida Flood Control District. (1954) *Five Years of Progress: 1949–1954*, p38

29. Central and Southern Florida Flood Control District. (1955) *Facts about F.C.D,*

30. Larsen, William F. (1954) *The Central and Southern Florida Flood Control Project*, Civic Administration Series, University of Florida Public Administration Clearing Service, Gainesville, FL pp21–22

31. Hollander, Gail M. (2008) *Raising Cane in the 'Glades*, University of Chicago Press, Chicago, p100

32. Nixon, Smiley. (1966) "Florida project mired in controversy," *Miami Herald*, 30 January

33. Central and Southern Florida Flood Control District. (1973) *23rd Annual Report of the Central and Southern Florida Flood Control District: July 1, 1971—June 30, 1972*

34. Worster, Donald. (1985) *Rivers of Empire*, Pantheon Books, New York, pp7, 276

35. Governor's Conference on Water Management in South Florida. (1971) "Statement to Governor Reubin O'D. Askew," *Water Management Bulletin*, vol. 5, no. 3

36. Central and Southern Flood Control District. (1975) *25th Annual Report*, p6

37. Oral History Program. (2005) Robert Clark, 8 July

38. South Florida Water Management District. (2009) *Kissimmee River Restoration: Just the Facts*

39. Oral History Program. (2005) Estus Whitfield, 29 August. Nat Reed, a Republican, a member of the SFWMD governing board and long-time Everglades activist, gave Graham unstinting praise for his efforts: "You cannot estimate it [Graham's contribution to the

Everglades restoration]. It is so great, it is so important that nothing that I could possibly say in the English language could come close to my strong feeling and my strong belief that we would never, ever, have gotten the Everglades Restoration Act signed into law without his continuing commitment." (Oral History Program interview of Nat Reed, November 2, 2000)

40. Rizzardi, Keith W. (2001) "Alligators and litigators: A recent history of Everglades regulation and litigation," *Florida Bar Journal*, vol. LXXV, no. 3

41. U.S. Army Corps of Engineers. (1989) *Central and Southern Florida Water Supply Study, Final Report*

42. Oral History Program. (2001) Interview of Dick Pettigrew, 23 May

43. Governor's Commission for a Sustainable South Florida. (1996) *A Conceptual Plan for the C&SF Project Restudy*, Governor's Commission for a Sustainable South Florida, Coral Gables, FL

44. U.S. Army Corps of Engineers. (1999) *Central and Southern Florida Project Comprehensive Review Study (The Restudy): Update and Background*

45. General Ballard: Chief of Engineers, Memo to Secretary of the Army, Central and Southern Florida Project – Comprehensive Review Study, June 22, 1999. Assistant Secretary of the Army Joseph Westphal: Letter to Vice President Albert Gore, July 1, 1999. Vice President Al Gore: Cited in Michael Grunwald (2006) *The Swamp*, Simon & Schuster, New York, p311

46. Grunwald, Michael. (2006) *The Swamp*, Simon & Schuster, New York, p344

47. Water Resources Development Act, 2000, Title VI, Section 601

48. Grunwald, Michael. (2002) "An environmental reversal of fortune," *Washington Post*, 26 June

49. General Accounting Office. (2007) *South Florida Ecosystem: Restoration Is Moving Forward but Is Facing Significant Delays, Implementation Challenges, and Rising Costs*. Report to the Committee on Transportation and Infrastructure, House of Representatives, General Accounting Office, Washington, DC

50. U.S. Sugar has expended more than $17.5 million on Florida election campaigns since 1976. Steve Bousquet and Jennifer Liberto. (2008) "Glades deal creates political void," *St. Petersburg Times*, 8 July

51. Morgan, Curtis, and Scott Hiassen. (2008) "Few knew details of sugar buyout talks," *Miami Herald*, 26 June

52. Quinlan, Paul. (2008) "Everglades at a crossroads as vote nears on U.S. Sugar deal," *Palm Beach Post*, 13 December

53. U.S. Sugar. (2008) "SFWMD approves historical deal to effect real and meaningful restoration of Florida's Everglades." Press statement, December 16, 2008

54. Hafenbrack, Josh. (2008) "Crist: God is happy after Everglades vote," *Lauderdale Sun-Sentinel*, 18 December

55. Paul Dumars, CFO of The PFM Group, in written communication to David Moore, South Florida Water Management District, February 17, 2010

56. Reid, Andy "Rick Scott gets rough reception at Tea Party protest over Crist's Everglades land deal" *Sentinel*, August 12, 2010

57. Barkett, John M. (2010) *Report of the Special Master (August 30, 2010)*, U.S. District Court, Southern District, 88-1886-CIV-Moreno

58. U.S. Environmental Protection Agency, "Amended Determination" for compliance with Judge Alan Gold's order of April 14, 2010. (Miccosukee Tribe of Indians of Florida, and *Friends of the Everglades v. United States of America, et al.*, No. 04-21488-GOLD/MCALILEY, and consolidated cases)

59. Pantano, Colonel Alfred A., Jr. (2009) "Everglades: The catalyst to combat the world's water crisis," Master's thesis, U.S. Army War College, Carlisle Barracks, PA

60. Morgan, Curtis, and Lesley Clark. (2009) "Stimulus aid for Glades," *Miami Herald*, 29 April

61. Matthew C. Godfrey and Theodore Catton provide an excellent review of how south Florida water policy has evolved, including their Chapter 18, "Getting the water right: The restudy and enactment of CERP, 1996–2000," in *River of Interests: Water Management in South Florida and the Everglades, 1948–2000* (2006) U.S. Army Corps of Engineers, Washington, DC. The Florida effort is put in context by Leonard Shabman, "Water resources management and the challenge of sustainability" in Roger A. Sedjo (ed). (2008) *Perspectives on Sustainable Resources in America*, Resources for the Future, Washington, DC, p121

62. Reid, Andy "South Florida Sets Three-Month Dry Weather Record" *Sentinel*, January 4, 2011.

PART III

THE ISSUES

CHAPTER TEN

Water Supply Is Easy

The concern about a supposed water supply "crisis" goes back to the origin of the modern water management system. The 1971 Model Water Code, the legal foundation for the Water Resources Act of 1972, sets the tone: "As a nation, the United States is in the early stages of a water *crisis.*" The (Florida) Governor's Conference on Water Management that same year picked up on the theme by advising Governor Reubin Askew in its first sentence that "There is a water *crisis* in South Florida today." A decade later in 1981, the South Florida Water Management District felt compelled to "take extraordinary measures" in another drought and ask the governor to declare a "*disaster emergency.*" The *Florida Bar Journal* published an article in 1993 pointing to the "Coming *Crisis* in Consumptive Use." A reporter for the *Orlando Sentinel* received a George Polk award for a 12-part series in 2002 titled, almost inevitably, "Florida's Water *Crisis.*" And in 2008, the annual report from the state's Century Commission for a Sustainable Florida concluded that "Our water supply is in *crisis.*"[1]

Can it really be true that a state with 54 inches of rain a year is in a perpetual water supply crisis? Or is it possible, instead, that the word "crisis" is used without full consideration of what it means and without considering achievable improvements in water use efficiency? Contrary to common perception, water supply in Florida is *not* in perpetual crisis. The water supply problems that do exist can be substantially resolved if already available and sustainable water conservation solutions are implemented. Thinking of Florida water supply problems as being in continual crisis has powerful adverse consequences. A crisis mentality discourages long-range thinking because a genuine crisis – when it exists – mandates short-term emergency responses. It prevents wasteful water supply projects from receiving the scrutiny they deserve, because action must be taken immediately in a crisis, regardless of harmful consequences. Water supply is relatively easy, even in a state with 18 million people, if the goal is enough water for sustainability for all rather than enough water for all to waste.

BILLIONS OF GALLONS A DAY

Water use data are not compiled annually by all of the water management districts but are tabulated at 5-year intervals by a cooperative agreement between the Florida Department of Environmental Protection and the U.S. Geological Survey.[2] (This critical data from the USGS is used throughout this chapter.) Freshwater withdrawals in Florida were 6.84 billion gallons a day in 2005. Saline water withdrawals were even larger, at 11.49 bgd. Despite high population growth rates in Florida, total withdrawals of freshwater have not increased much since 1975 in any water management district, with the exception of the South Florida Water Management District (see Figure 10.1)

Floridian anxiety about water supply (and water quality) should be understood in a global context. At the same time that residents worry about securing enough water to overirrigate their St. Augustine grass or fret about which brand of bottled water to buy, about 1.1 billion people across the globe lack safe drinking water and 2.6 billion go without even the most basic sanitation facilities. Five thousand children around the world die each day from easily preventable water-borne diarrheal diseases.[3] Florida has nothing like this magnitude of water-related problems.

WITHDRAWAL VERSUS CONSUMPTION

Water is "withdrawn" when it is removed from a ground or surface source for human use. A municipal drinking water well may withdraw, for example, a million gallons of water a day and transport that to a drinking water treatment plant for delivery to customers. In contrast, water that is "consumed" is evaporated, transpired by plants, or otherwise made unavailable to the source from which it has been withdrawn. The amount withdrawn is the number almost often reported,

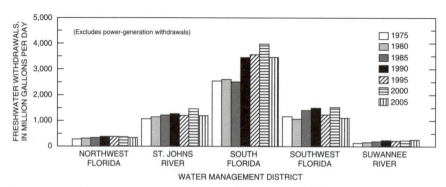

Figure 10.1 *Freshwater withdrawals in Florida, 1975–2005*

Source: U.S. Geological Survey

but it can be much less important than the amount consumed because consumption is what actually depletes the resource.

For example, a power plant using water for cooling may withdraw a hundred million gallons of water a day but return almost all of that to the source. The consumption of water, in this case, is only a tiny fraction of the withdrawal. Irrigation is one of the activities with the highest fraction of consumed water. If the irrigation method is especially efficient – applying only the exact amount of water that the plants need – then a much higher fraction of the applied water is "consumed" in the irrigation process rather than running off or soaking into the ground. Almost all of the water applied to the plants evaporates or is used in plant respiration with "efficient" irrigation. In 2005, about 40 percent of the water withdrawn in Florida was consumed. The largest consumption of any sector was in agriculture.

THE PATTERNS OF WATER USE

Water is not withdrawn equally across the state. About half (1,538 mgd) of all water utility supplies are withdrawn in a single water management district – South Florida – which has both the largest population of any district and the largest center of agricultural irrigation. To understand other patterns of water use, the types of water users have to be evaluated separately. The U.S. Geological Survey (USGS) categorizes water use into six sectors and provides estimates both of water withdrawn and water consumed. Table 10.1 provides the basic data on how much water is withdrawn by different types of use:

1. *Public Supply* (water utilities that deliver drinking water to customers).
2. *Domestic Self-Supplied* (individuals on private wells plus people served by very small utilities that pump less than 10,000 gallons a day. About 96 percent of the domestic self-supplied category is from private domestic wells.)
3. *Commercial-Industrial Self-Supplied* (schools, prisons, hotels, and other institutional facilities, as well as other commercial and industrial activities such as mining and manufacturing).

Table 10.1 *Water Withdrawals in Florida, 2005*

	Fresh Water			Saline Water		
	Ground	Surface	Total	Ground	Surface	Total
Public supply	2201	339	2,541	0	0	0
Domestic self-supplied	186	0	186	0	0	0
Commercial-industrial self-supplied	366	123	488	0	1	1
Agricultural self-supplied	1,302	1,465	2,766	0	0	0
Recreational irrigation	171	159	330	0	0	0
Power generation	18	541	558	3	11,481	11,484
TOTALS	4,242	2,626	6,868	3	11,482	11,486

Source: Compiled by the U.S. Geological Survey, Tallahassee; all values in million gallons per day

4. *Agricultural Self-Supplied* (including irrigation and nonirrigation uses. Almost all of the water use in this sector is applied to crops but some is used for livestock watering, fish farming, and other agricultural uses.)
5. *Recreational Irrigation* (landscape irrigation used in recreational activities such as golf courses and ballfields.)
6. *Power Generation* (use of water by power plants for cooling, either in once-through systems or as make-up water for recirculating systems.)

PUBLIC WATER SUPPLY

Total withdrawals for Florida public water supply were 2.54 billion gallons a day in 2005. It is a governmental activity: more than 90 percent of water from public water suppliers is delivered by government utilities rather than privately owned. A subset of customers, sometimes derided as "water hogs," use a disproportionate amount of water. In Orlando, for example, more than 10,000 homeowners each use more than 20,000 gallons a month.[4]

Public water suppliers in Florida rely mostly on fresh groundwater for their supplies. However, a growing (but still small) fraction of public water supply is derived from brackish groundwater. Florida has more than 130 energy-intensive demineralization facilities — more than any other state.[5]

How many gallons per capita are used in Florida each day? The answer is surprisingly hard to obtain because estimates of both the gallons and the people in the "per capita" formulation have some definitional and empirical uncertainties. The USGS estimates, with their methods, that the public supply "gross" per capita water use in 2005 was 158 gpcd. According to the USGS, this was "lower than any previous year, and reflects a combination of normal precipitation for 2005 and the effects of water conservation and restrictions imposed over the past 10 years." This trend is good news because it points to achievements in improving water use efficiency.

To estimate the separate *residential* per capita use of water, the USGS deducts from public water supply totals the amount they estimate was used by other water use sectors, then divides by the same estimate of population they used in calculating gross per capita use of water. Their residential estimate is always lower than the estimate of total per capita because it excludes significant water users such as businesses. The USGS estimate of residential water use in Florida in 2005 is 95 gpcd. This also was lower than any previous estimated residential per capita use. They attribute the apparent reduction in residential per capita use to:

- Implementation of water conservation measures in the last decade
- Substitution of reclaimed domestic wastewater for potable water
- The spread of Florida-Friendly landscaping.

Up to half, or even more, of the municipal water supply is not used for drinking water, for cooking, for sanitation, or for any other indispensable application.

In some communities about 50 percent of the water treated to drinking water standards by water utilities ends up being used for landscape irrigation. This extraordinary waste of drinking water for landscape irrigation is exacerbated by the common use of far more water than the plants need.[6] This matters because Florida has about 4470 square miles of turf (lawn grass). This turf occupies 7.1 percent of the land mass of the state, which is more than three times higher than the national average of 1.9 percent.

Trend in Public Water Supply

The USGS reports that total public water supply withdrawals increased 32 percent between 1990 and 2005, while population increased in Florida by 44 percent. The good news is that urban demand for water increased less than population did. The not-so-good news is that water use in this sector increased significantly. Over the 1990–2005 period, public water supply demands increased more rapidly than any other water use sector.

DOMESTIC SELF-SUPPLIED WATER

Homes not on public water supply systems withdrew about 185 mgd in 2005 for about 1.75 million people. The number of people on domestic wells is about the same today as it was in 1980 and is about 10 percent of the population. The USGS estimates that the per capita use in the domestic category was 95 gallons per day in 2005 – lower than any year in the past. However, domestic use from private wells is not metered or reported to any agency, so the estimates are subject to substantial uncertainty.

Trend in Domestic Self-Supply

The USGS estimates that demand for water by homes not on a community water system decreased almost 40 percent between 1990 and 2005. The decrease in estimated water use may not be real, but due to the adoption by USGS of more accurate methods of estimating water use in this sector.

COMMERCIAL-INDUSTRIAL SELF-SUPPLIED WATER

This water use sector is for activities that have their own water withdrawal facilities and also those that purchase water from a public utility. This sector includes a diverse set of water-using activities, from mining to industries to schools, prisons, and commercial businesses such as hotels and resorts. Mining at 70 locations across the state was responsible for 40 percent of withdrawals in this water use sector in 2005.

Self-supplied withdrawals in 2005 totaled 488 mgd, mostly from groundwater. An even larger amount of 591 mgd was purchased by users in this sector from public water suppliers. The amount of water supplied to this category of users from public water supply systems increased 27 percent (126 mgd) from 2000 to 2005.

Cooling towers, a part of the air conditioning systems in commercial buildings, are a significant use of water in Florida's climate. Fifteen percent of the nation's 90,000 U.S. cooling towers are in Florida, with Miami-Dade County holding the title of "Cooling Tower Capital of the World."[7] There are significant opportunities to reduce the amount of water used in cooling towers with new technologies that allow the water to be recycled several times before being discharged.

Trend in Commercial-Industrial Self-Supplied

The amount of water used in this sector by self-suppliers decreased by 36 percent between 1990 and 2005. On the other hand, commercial-industrial water users that purchased water from public water suppliers increased by 27 percent over the same period. Combining both types of water use, the amount was about the same at the end of the 15-year period, despite a large increase in population and economic activity.

AGRICULTURAL WATER USE

Florida uses more water in agriculture than any other state east of the Mississippi River. Farms, groves, and ranches withdrew about 2.77 bgd for this purpose in 2005. The Everglades Agricultural Area, in particular, uses enormous amounts of water, largely to grow subsidized sugarcane. The EAA alone accounts for 40 percent of surface water withdrawals in the South Florida Water Management District.

Overall, agriculture is the second largest category of water use in the state. Only the public water supply sector (37% of withdrawals) was larger in 2005. A very different picture is seen in terms of agricultural consumption, however. Agriculture consumes about three-fifths of withdrawals. As a result of these higher rates of consumption of water, agriculture consumption amounted to more than twice that of the 28% of the total water consumed in public supply.

Aside from almost 2 million acres of pasture, most crops in Florida are irrigated. Agricultural irrigation demands are highest when water is in shortest supply during a drought. Of the 2.02 million acres of farmland other than pasture, about 1.67 million acres were irrigated in 2005. The average irrigation amounts for some representative "crops" in Florida are shown in Table 10.2.[8]

Trend in Agricultural Irrigation

The USGS data show a marked reduction in irrigation amounts from 1990 to 2005 and an even larger percent reduction in the 2000–2005 period. From 1990–2005, agricultural irrigation decreased by 21 percent and, from 2000–2005, the decrease was nearly 30 percent. Part of this reduction is due to a change in how irrigation

Table 10.2 *Crops and Water Use*

Land Use	Average Water Use (Gallons per Acre Per Day)
Corn	1135
Watermelon	1104
Sugar	2195
Citrus	2415
Sod	2474
Tomatoes	2666
Container-grown ornamentals	4912

amounts are estimated by the USGS (much of it is not measured accurately or reported regularly to the water management district). However, 2005 also was wetter than 2000 and the total amount of agricultural acreage in Florida declined over that period.

Chapter 13, Farming the Water, provides more details on Florida agriculture's use of water.

RECREATIONAL IRRIGATION

Floridians use so much water that it does appear at times that irrigation is a recreational activity all by itself. However, what the USGS means by "recreational" irrigation actually refers to applying irrigation water to recreation sites or for "aesthetic" purposes. About 331 mgd of fresh water was used for this purpose in 2005, plus almost that same amount of reclaimed water. Florida has more golf courses than any other state and close to 90 percent of course acreage is irrigated.[9] The amount of water used on golf courses increased 17 percent between 1990 and 2005.

Trend in Recreational Irrigation

Withdrawals in 2005 for this purpose were 17 percent higher than in 1990, but also were 20 percent lower in 2005 than in 2000.

POWER GENERATION WATER USE

Almost all (95 percent) of the water used in thermoelectric plants in Florida is saltwater. Only California withdraws more saline water than Florida. This use of oceanic sources for power plant cooling is one of the successes of Florida water policy because it preserves potable freshwater mostly for uses that actually need high-quality water. Water consumption in this sector is small because 99 percent of the water withdrawn is returned directly to the source.

Trend in Power Generation

The amount of water used for this purpose decreased 29 percent between 1990 and 2005. Looking only at the period 2000–2005, water use decreased by 15 percent, even though power generation increased.

OVERALL TRENDS

Nationally, freshwater withdrawals peaked in 1980. For 2005, the most recent year with national data available, total withdrawals in the United States were similar to 1990 even though population increased 13 percent over that period. From 1985 to 2005, the amount withdrawn has varied less than 3 percent.[10] Florida, taking into account that its rate of population growth has been higher than for the country as a whole, has followed about the same trend.

One reason for demand not growing very much is the success of Florida's program to reuse domestic wastewater. Of the 1,451 mgd of domestic wastewater treated in Florida at plants with larger than 100,000 gallons per day capacity, 667 mgd (46 percent) is put to some beneficial reuse, such as landscape irrigation. That significantly reduces demand for potable supplies from groundwater and other sources.[11]

Population in Florida increased by 44 percent between 1990 and 2005, while public water supply withdrawals increased only 32 percent. Taking a broader look at all uses of water combined, total fresh water withdrawals actually *decreased* by 715 mgd (9 percent) between 1990 and 2005, while the population increased by 1.9 million people. From 2000 to 2005, total withdrawals of fresh water continued to *decrease*, this time by 1,324 mgd. (The trend of decrease from 1990 to 2005 is clear, but the 2000–2005 period bridges a relatively wet period in 2000 and a relatively dry period in 2005.)

ALTERNATIVE WATER SUPPLIES

Old-style impoundment of rivers has fallen out of fashion in Florida. So has draining lakes and wetlands by overpumping of wellfields, as happened in northern Tampa Bay in the 1990s. What is not yet in fashion is full efficiency in water use. "Alternative" supplies of water are being sought everywhere. This search for nontraditional sources of water is often called "growing the water pie."[12] Indeed, the founder of the Sugar Cane Cooperative of Florida believes that "the greatest thing coming out of the Comprehensive Everglades Restoration Program [CERP] is its focus on water quantity. It envisions enlarging the water pie to avoid the water wars that have been so destructive in the West."[13] The director of governmental affairs for the Florida Fruit and Vegetable Association agrees: "Any time they make the water pie bigger, it means more water for agriculture."[14]

Floridians often try to eat a "water pie" with a fork and drop most of the water before it reaches the mouth. The problem is that the pieces of the water pie are expensive. Florida has several scores of plants (large and small) to remove salts from brackish groundwater and seawater.[15] Water supply projects are very vulnerable to cost inflation and design shortcomings. Capital facilities for water supply usually cost far more than first estimates by proponents. A national study of infrastructure projects found that "legislators, administrators, bankers, media representatives, and members of the public who value honest numbers should not trust the cost estimates presented by infrastructure promoters and forecasters."[16]

In practice, "alternative" water is often synonymous with the word "costly," and water users often turn to the water management district or the state legislature in their quest for dollars. In 1997, the legislature stated that water management

Figure 10.2 *Key West depends on a 130 mile pipeline running from the Florida mainland along a chain of islands and bridges*

districts were to "take the lead" in water resource development projects and that local governments and utilities were to "take the lead" in water supply development projects. (Section 373.0831(2)(c), F.S.) The legislature provided a clear policy that water users should receive a clear price signal:

> Generally, direct beneficiaries of water supply development projects should pay the costs of the projects from which they benefit, and water supply development projects should continue to be paid for through local funding sources. (Section 373.0361(2)(c), F.S.)

That principle of "user pays" held as a general principle until 2005. Water users – principally public water suppliers – lobbied the legislature to help pay for their water supply development projects. In response, the legislature created the Water Protection and Sustainability Program. The state provided funds on a cost-share basis for local water supply projects as long as the projects were for "alternative" water supply development projects. That term was interpreted to include projects to treat brackish groundwater to drinking water standards, upgrading reclaimed water facilities, and the construction of conventional groundwater wells in north Florida. The drive to develop "alternative" water supplies at state expense may be only the latest iteration of expensive capital facility projects in Florida's water history.

The state and national fiscal crisis forced the state legislature to first reduce and then eliminate all funding for the Water Protection and Sustainability Trust Fund by 2009.

WATER IN LITTLE BOTTLES

Bottled water is extremely costly, considering that tap water costs only a fraction of a cent for a gallon physically delivered to a home and doesn't have to be picked up by hand in the store and then transported home. It surprises some purchasers that bottled water is not subject to the rigorous monitoring requirements for water delivered from utilities.

Among the mysteries of the appeal of bottled water is the special cachet of drinking so-called spring water. The U.S. Food and Drug Administration allows labels to claim the source is from springs, but does not require that the point of withdrawal be from the spring itself. It is only necessary that the well derive the water from the "same underground stratum" as the spring, have some "measurable hydraulic connection," and have the same "composition and quality."[17] Despite some consumers' preference for "spring" water, about 44 percent of bottled water sales in the United States derive from regular tap water.[18] Even "spring" water does not always carry enough supposed purity for many consumers. Some water bottle labels explain that the container holds only very pure water produced by an energy-intensive reverse osmosis process. This often is done merely to re-clean safe tap water from regular municipal supplies.

The Pacific Institute has provided some useful calculations on the resources involved in bottling water. They estimate that 17 million barrels of oil are used annually in manufacturing the bottles – equivalent to filling each bottle one-quarter with oil.[19] One thousand 12-ounce water bottles, on a lifecycle basis, are associated with the emission of 9.6 kg of carbon dioxide.[20] The state of Oregon concluded that the global warming effects of bottled water were 46 times larger than for tap water.[21]

Bottled water is not a large consumer of water itself (even if it comes in very wasteful plastic bottles). The USGS estimates that water bottlers withdrew about 3.6 mgd in Florida but an additional unknown amount came from public suppliers.[22] Exact numbers are proprietary and the legislature has not required that they be reported to anyone. As in other states, water bottling operations in Florida today are often very controversial, including battles in the Florida legal system. In 2008, the Fifth District Court of Appeals upheld by 2–1 the issuance of a water bottling permit from the St. Johns River Water Management District for a well in Marion County, but only grudgingly.[23] One of the two judges in the majority did not "think it is a good idea to allow this consumptive use of fresh water resources" but followed what he believed the statute authorized.

From one perspective, the few millions of gallons of water a day allocated to water bottling in Florida do not merit a high level of concern. It is not a significant fraction of water use in Florida. Each bottling withdrawal permit has to pass the same permitting test as any other water use. Bottled water is simply another item of luxury convenience, like sports drinks and coffee latte, purchased in a wealthy nation. It is a choice, like others, which causes more material consumption, more energy use, and more emission of greenhouse gases. It is no worse, and no better, than many other water management choices.

CONCLUSION

Nelson Blake, before he wrote his classic 1980 book on the history of water management in Florida, published another book in 1956 on the national urban water supply experience. In his view, "The municipal statesmen and engineers who preserve and enrich life through providing an essential service deserve to have their names and deeds recorded in the history of the nation no less than do presidents, generals, and philosophers."[24] Florida water utilities, civil engineers, water managers, and local governments have put in place a very reliable water supply system, for which they deserve great appreciation. However, it is time to rethink the fundamental skills that have made billions of gallons of water a day available in Florida. New approaches are needed, including ones that emphasize water use efficiency over water supply development. With improved efficiency, there would be much less talk of a water supply "crisis." That potential future is the subject of the next chapter.

NOTES

1. Various sources include: Model Water Code, p v; Governor's Conference on Water Management in South Florida, 1981; Robert Clark, Jr., Chair, SFWMD Governing Board,

letter to Governor Bob Graham, July 16, 1981; Richard Hamann and Thomas T. Ankersen, "Water, wetlands, and wildlife: The coming crisis in consumptive use," *Florida Bar Journal*, March 1993, pp41–46; Debbie Salamone, "Florida's water crisis," *Orlando Sentinel*, 2002 series; Century Commission for a Sustainable Florida, *Second Annual Report to the Governor and Legislature*, January 16, 2008, p7.

2. Marella, 2009.

3. World Health Organization (WHO). (2005) *Water for Life*, WHO and UNICEF, pp1–33

4. Spear, Kevin. (2008) "Water hogs: Central Florida's celebrities, average Joes waste supply. Celebs lead the list, but average Joes waste, too," *Orlando Sentinel*, 7 September

5. Florida Department of Environmental Protection. (2010) *Desalination in Florida: Technology, Implementation, and Environmental Issues*

6. Haley, Melissa B., Michael D. Dukes, Stacia Davis, Mary Shedd, and Bernard Cardenas-Lailhacar. (2008) Energy efficient homes: The irrigation system," Document FCS3274, Florida Cooperative Extension Service, IFAS, University of Florida, http://edis.ifas.ufl.edu/FY1043, accessed 14 August 2010

7. South Florida Water Management District. (2008) *Water Conservation: A Comprehensive Program for South Florida*, p23

8. Obreza, Thomas. (2004) *Florida Springs Land Use Information Tool*, Circular 1448, Soil and Water Science Department, Institute of Food and Agricultural Sciences, University of Florida, Gainesville, FL

9. Haydu, John J., Alan W. Hodges, and Charles R. Hall. (2006) *Economic Impacts of the Turfgrass and Lawncare Industry in the United States*, EDIS Document FE632, Institute of Food and Agricultural Sciences, University of Florida, Gainesvelle, FL, p24

10. Kenny, J. F., N. L. Barber, S. S. Hutson, K. S. Linsey, J. K. Lovelace, and M. A. Maupin. (2009) *Estimated Use of Water in the United States in 2005*, U.S. Geological Survey Circular 1344, U.S. Geological Survey, Washington, DC

11. Florida Department of Environmental Protection. (2010) *Water Reuse Program, 2008 Reuse Inventory*

12. Northwest Florida Water Management District. (2007) *Consolidated Annual Report, March 1, 2007*, p109

13. Oral History Program. (2002) George Wedgworth, 25 April

14. Florida Fruit and Vegetable Association. (2008) "Governor Crist supports alternative water supply funding," *Rap-Up Newsletter*, issue 1208

15. Florida Department of Environmental Protection. (2010) *Desalination in Florida: Technology, Implementation, and Environmental Issues*

16. Flyvbjerg, Bent, Metter Skamris Holm, and Soren Buhl. (2002) "Underestimating costs in public works projects: Error or lie?" *Journal of the American Planning Association*, vol. 68, no. 3, p291. The focus of this article was on transportation infrastructure but they also note, "In addition to cost data for transportation infrastructure projects, we have reviewed cost data for several hundred other projects including power plants, dams, water distribution, oil and gas extraction, information technology systems, aerospace systems, and weapons systems. The data indicate that other types of projects are at least as, if not more, prone to cost underestimation as are transportation infrastructure projects." (p286)

17. U.S. Food and Drug Administration, 21 CFR Ch. I (4-1-03 ed), Subpart B – Requirements for Specific Standardized Beverages, § 165.110 Bottled Water

18. Royte, Elizabeth. (2008) *Bottlemania: How Water Went on Sale and Why We Bought It*, Bloomsbury, London, p38

19. Pacific Institute. (2010) "Bottled water and energy fact sheet," www.pacinst.org/topics/water_and_sustainability/bottled_water/bottled_water_and_energy. html, accessed 14 August 2010

20. Franklin Associates [division of ERG]. (2006) *Life Cycle Inventory of Five Products Produced from Polylactide (PLA) and Petroleum-Based Resins, Summary Report.* Prepared for Athena Institute International, pp1–21

21. Franklin Associates [division of ERG]. (2009) *Life Cycle Assessment of Drinking Water Systems: Bottle Water, Tap Water, and Home/Office Delivered Water.* Revised final peer-reviewed LCA report, prepared for the Oregon Department of Environmental Quality, October 22, 2009

22. Marella, op. cit., note 2

23. *Marion County v. C. Ray Greene, III and Angus S. Hastings*, 1st D.C.A., Florida, Case No. 5D07-1239, July 2008

24. Blake, Nelson. (1956) *Water for the Cities: A History of the Urban Water Supply Problem in the Unites States*, Syracuse University Press, Syracuse, NY, p2

CHAPTER ELEVEN

Saving the Water

D eveloping new water supplies in Florida is like putting a larger air conditioner on a house while leaving the front door wide open. Water efficiency is even more valuable than energy efficiency in a hot and humid state.

The state is committed, at least in general, to water conservation. For example, a 1985 newspaper editorial commended the governing board of the South Florida Water Management District for meeting in special session and agreeing that "from now on the message to the district's four million people in 16 counties and the multimillion dollar agricultural industry is permanent conservation." District board members were quoted as saying that "the days of plentiful and cheap water are over." The new philosophy was to "reduce demand before any future water supply projects are planned."[1] However, more than two decades later, the South Florida Water Management District, as well as the rest of the state's water management system, is still struggling to develop an effective water conservation program. In 2008, the chair of the South Florida District's governing board described the district's "new" water conservation program in terms very similar to those used in 1985. The chair noted that the South Florida District "is at a critical turning point," that the "days of cheap and unlimited water are over," and that water conservation has the "very real potential to defer or reduce the need for more costly development of new sources and facilities."[2]

Although there are important water conservation programs in Florida, the biggest and cheapest source of water for the state's future remains water that is now wasted. More effective water conservation measures could meet a large fraction of Florida's water supply needs. Achieving these potential improvements would be long steps on the path to water resource sustainability.

Although water conservation opportunities remain, Florida has implemented a number of innovative programs to improve water use efficiency. For example, water use permitting programs of the water management districts explicitly require applicants to implement at least some efficiency measures. These and other water conservation programs around the state have helped achieve some notable successes. Some regions and communities have achieved more than others:

- Per capita use of public water supply in the Southwest Florida Water Management District declined 19 percent since the early 1990s, from 142 gpcd (the average for calendar years 1991 and 1992) to an average of 115.5 (for calendar years 2006 and 2007). Some counties in the Southwest Florida District have done even better. Pinellas County has seen a 24 percent reduction over that period. Sarasota achieved a 37 percent reduction (from 133 gpcd in the early 1990s to an average of 84.5 gpcd in 2006 and 2007).
- The member governments of Tampa Bay Water have saved 26 mgd of water since 1996 and they expect to save up to 36 mgd total by 2012.[3] These measures are an integral part of their water supply plan for the future.
- The city of Oviedo (near Orlando) made free water audits available to residents under the "H2Oviedo Water Conservation Incentive Program." Homeowners receive up to $1,000 toward the costs of installing "Florida-Friendly" landscaping. Customers whose water use is 10 percent or more below the amount budgeted for their home receive a $50 bonus.[4]
- In 2008 the St. Johns River Water Management District adopted a rule that limits landscape irrigation to only 2 days a week during the Daylight Savings Time part of the year and to only 1 day a week during Standard Time. (The winter season, when irrigation can occur only 1 day a week, is when the water needs of plants are the lowest.) They also restrict each irrigation event to no more than an hour and to applying no more than three-quarters of an inch of water.
- The Miami-Dade Water and Sewer Department developed a water conservation plan intended to save at least 19.6 mgd.[5] The actual use of water appears to have exceeded their objectives. Miami-Dade water conservation efforts, combined with water shortage orders of the South Florida Water Management District, reduced demand for water from 340 mgd in 2006 to less than 310 mgd in 2008. Per capita use of water decreased from about 157 gpd in the 2002–2005 period to an average of 139 gpd in 2007–2008.[6]

Although significant, these accomplishments are less than those achieved elsewhere. For example, San Antonio, Texas reduced the per capita use of water almost a third from its level in the early 1980s. The "Saving Water Partnership" of Seattle and 17 other utilities in that region have reduced water use by 44 mgd (26 percent) from 1990 to 2007, while population increased 16 percent over that period. The region has also set a goal of an additional 15 mgd in annual savings by 2030. The population of Los Angeles grew by 33 percent between 1975 and 2005 but without increasing water use. Las Vegas managed to achieve a 13 percent reduction in per capita water use from 2003 to 2005. An ordinance requiring the use of rainwater catchment systems has been put in place by Tucson, Arizona. Beginning in 2010, new developments must met 50 percent of their landscaping requirements by capturing rainwater on-site. [7]

Using less water, rather than developing new water supplies, has many advantages, such as:

- Preventing the destruction of water resources, because new water supply projects are not necessary after improving efficiency in the use of existing supplies.[8]
- Facilitating the construction of necessary water supply projects by reducing public doubts about the genuine need for them.
- Saving money on constructing water supply projects, because the need to build expensive new facilities can be eliminated or deferred.
- Saving money on the costs of operation and maintenance, because water efficiency measures are often cheaper than capital facilities such as reservoirs and seawater desalination plants.
- Preventing harm to water bodies from overwithdrawals by making it easier to maintain adopted minimum flows and levels.
- Reducing a community's emissions of greenhouse gases, because water conservation requires less energy than capital facilities for water supply projects.
- Responding quickly to short-term water supply needs, because water efficiency measures can be implemented in a shorter time than large construction projects.
- Diversifying the "sources" of water so that a community is not relying upon only one, or even a few, sources of water for its water future.
- Providing reliable long-term savings, because many water conservation measures are technology-based and certified to perform well.
- Giving individuals more control over their water bills rather than requiring all of them to share involuntarily in the cost of long-term capital facilities.
- Reducing flow to wastewater treatment plants, thereby reducing the associated costs of treatment.
- Reducing vulnerability to drought by using less water when rainfall is deficient.

Notwithstanding these many advantages, water conservation programs in Florida are falling far short of their potential. The potential is known because it has been studied many times, but with only partial implementation after each study was completed. The state's response to two recent droughts demonstrates lost opportunities. (See Table 11.1 for a listing of recent efforts.)

Interest in water conservation increased during the 1999–2001 drought. In the midst of that drought, the Florida Department of Environmental Protection led a "Water Conservation Initiative" that produced 51 separate recommendations in 2002

Table 11.1 *Recent Water Conservation Reports and Events*

2002	Water Conservation Initiative
2003	State law requires a "comprehensive program" for water conservation in public water supply.
2004	Conserve Florida Clearinghouse established at the University of Florida.
2006	"Being Drought Smart" report released.
2008	Florida section of the American Water Works Association 2030 recommendations released.
2008	South Florida Water Management District approves a new comprehensive water conservation program.
2008	Commission for a Sustainable Florida organizes a "Water Congress" meeting.
2008	Governor's Energy and Climate Change Action Team produces final report.

for water use efficiency.[9] Hundreds of concerned people and water management professionals collaborated in this effort, showing strong public support. However, only one-third of those recommendations were implemented even a decade later.

The Water Conservation Initiative did produce a written plan on how to improve water use efficiency in the public water supply sector. In February 2004, a "Joint Statement of Commitment for the Development and Implementation of a Statewide Comprehensive Water Conservation Program for Public Water Supply" was signed: the Florida Department of Environmental Protection, the five water management districts, and the public water supply industry (via the Florida section of the American Water Works Associations, the Florida Water Environment Association, and the Florida Rural Water Association) agreed on key principles of how to promote efficiency. Once the relevant agencies and utility organizations had agreed on the basic principles in the Joint Statement of Commitment, it proved relatively easy to persuade legislators to amend the Florida Water Resources Act along the same lines. The state Department of Environmental Protection was directed to develop "goal-based, accountable, tailored, and measurable water conservation programs for public water supply" (Section 373.227, F.S.).

The legislature somehow forgot to appropriate any money to implement the new program that they had just mandated to be established. To start the program, FDEP had to solicit voluntary funding from both the water management districts and the utility sector. The program is now called Conserve Florida and operates primarily through a clearinghouse at the University of Florida. It provides a free statewide service for water utilities in planning and monitoring water conservation programs. FDEP acts as contract manager for the signatories of the Joint Statement. The clearinghouse continues to be funded on an ad hoc basis.

Another Florida drought came in 2006–2008. This second statewide drought of the twenty-first century occurred without the state having implemented many of the recommendations written at the end of the previous drought. At the state level, FDEP again drew together concerned agencies and organizations to make recommendations on measures that would both improve the immediate response to the drought and make long-term water conservation improvements. The resulting July 2007 Drought Smart report had 25 separate action recommendations, each assessed separately for drought responsiveness and water conservation effectiveness.[10] Just as for the previous Water Conservation Initiative, implementation has been very limited. By 2011, less than a fifth of the recommendations were even partly implemented by water management agencies and water suppliers.

In 2007, the state water utility organization (the Florida section of the American Water Works Association) undertook another significant water conservation effort outside the framework of FDEP and the water management districts. They began a Florida 2030 vision and planning exercise, with recommendations the next year, such as:

- All classes of water use will be at the highest feasible level of water use efficiency.
- Water conservation will be the priority water supply option considered to reduce new demands, and ranked for implementation based upon its benefit and cost effectiveness.

- All water users, except for domestic uses and minor agricultural activities, will measure and report their water use regularly to the water management districts.
- Per capita use in urban areas will be significantly less than today.[11]

Time will tell if the "vision" of 2030 is a meaningful goal or becomes a forgotten dream, as some previous visioning exercises turned out to be.

Another important milestone was the September 2008 Water Congress meeting described in Chapter 3, The Responsibilities of Water Management. Two of the top four recommendations from the Water Congress were for measures to improve water use efficiency:

- Amend, as necessary, any statute, rule or policy so that quantifiable water conservation best management practices are considered an "alternative water supply" and are equally as eligible for funding as capital facility expansion proposals.
- Set a per capita target or goal for water use and quantifiable best management water practices and provide a stable funding base for the Conserve Florida program directed by Section 373.227, F.S., including the statewide water conservation clearinghouse for public water supply.

Lastly, in October 2008, the Governor's Action Team on Energy and Climate Change delivered their final report with recommendations on how Florida should respond to global climate change. They recognized the connection between water use efficiency, greenhouse gases, and climate change. To reduce the greenhouse gas emissions associated with water supply activities, the Action Team recommended "intense conservation of all uses" and a number of specific water conservation actions.[12]

It is clear, from all of these recent reports, that there is strong support, at least in principle, for water conservation. Water conservation always sounds good. Progress, however, has been very uneven. The recommendations below are based on the consensus found in the many recent reports.

THE NEED FOR STATE LEADERSHIP

The need for action begins at the state level. The decentralized nature of Florida's water management system can make it hard for the state to assert leadership. As a result, the state often defers to the water management districts, or local interests, because they are closer to the problems of their regions (and are often better funded). However, there also are times when the state could take appropriate actions that would make a large difference.

Require the Best Available Water Efficient Technologies

The state could direct the use of tested and proven water-conserving technologies, such as EPA WaterSense-labeled products. These certified products use less water than other devices while attaining excellent customer satisfaction ratings. This option was one of the specific recommendations, for example, from the Florida 2030 vision process. The state could accomplish this goal by a vote of the Florida Building Commission to change the state building code. Alternatively, the legislature could enact the same policy. The state of Georgia did so in early 2010.

Treat Water Conservation Equally in Funding Programs

The idea of treating water conservation equally with supply projects is not new. In 1999, FDEP called on the water management districts in their regional water supply plans to treat water conservation "in the same manner, and to the same level of detail, as alternatives to create new supplies."[13] The 2005 Florida legislature chose a different path when it created the Water Protection and Sustainability Program. That law allowed state funding to be spent only on construction of capital projects. Projects to improve water use efficiency were not eligible for the financial assistance. In the first two years of the Alternative Water Supply Program, the state provided $160 million toward construction costs for capital projects, but no money was allocated to reduce the demand for water which itself generated the call for more water supply projects. This bias against efficiency does not save money because water conservation projects often cost 20 to 80 percent less than capital-intensive water supply projects.[14]

It would be much more effective to treat water conservation equally in the funding of water supply activities. Local interests should be allowed to decide for themselves whether they would rather be more water efficient or build new capital facilities when using state dollars. Some small progress in this direction was made by the 2010 legislature when it made water conservation projects eligible for equal funding in at least one program – the Water Protection and Sustainability Trust Fund. The legislature appropriated no money, however, for the fund that year.

Provide Adequate Funding for the Conserve Florida Clearinghouse

The Conserve Florida Clearinghouse is the principal state body promoting water conservation in public water supply. However, it has no dedicated source of funding. Its budget is much less than 1 percent of statewide expenditures on water supply development. The state should find a way to put the Clearinghouse on a stable funding base. For example, the legislature could dedicate a percentage of the state Drinking Water Program toward the Clearinghouse. With stable and adequate funding, the Clearinghouse could become a more effective institution and assist water utilities in optimizing their water conservation programs.

Facilitate the Use of Cisterns and Gray Water

Cisterns can collect substantial amounts of rain for nonpotable uses. A half-inch of rain falling on a roof 2,000 square feet in size can capture about 500 gallons of water. Keeping the water on-site for later use would reduce stress on public water supply systems.

Similar opportunities may exist for gray water, which refers to select sources of water in the home like that from clothes washers, bathtubs, and showers. Although not safe for drinking, gray water is relatively clean and does not include water from toilets. With adequate care, it could be used safely for other household uses such as toilet flushing and landscape irrigation. The legislature, or the Florida Department of Health (which regulates these activities), could take an active role in researching and/or promoting the safe use of rain cisterns and gray water.

FIX LANDSCAPE IRRIGATION

When the Governor's Conference on Water Management in South Florida was held in 1971, relatively few homes were built with underground networks of pipes to deliver water to lawns. Today, most new homes built in Florida come with a pre-installed technology to use and waste enormous quantities of water: an automatic in-ground irrigation system. The developer or builder delivers to the purchaser a home with an automatic irrigation controller hung on the wall in the garage. The controller manages complex settings for different irrigation zones, the amount of water to be applied, as well as the time of day and the days of the week in which water is to be drawn from the local water source and placed on the landscaping. Homebuilders do not sell the irrigation systems as an add-on option; the central systems are installed as part of the basic house package.

The problem is that homes with even simple in-ground irrigation systems use one-third more water than homes without them.[15] For homes with such devices, the owner does not even have to remember to turn on the system. The irrigation system controller turns on the sprinklers automatically without any decision by the resident. Homes with fully automatic irrigation systems use 47 percent more water than homes without them. In central Florida, up to two-thirds of residential water use is in automatic irrigation systems. Some systems use two or even three times as much water as the landscaping needs. In effect, the controllers are out of control.

Automatic irrigation systems practically guarantee automatic use of enormous amounts of water. About half, or even more, of all the water use at homes with such systems is used for irrigation, mostly in amounts far above what is required for the landscaping. The easiest way for homeowners to save water in homes is to not install an in-ground irrigation system. The homeowner would then save many hundreds or thousands of dollars on the cost of installing the system, probably could put in less expensive landscaping that does not require as much water, and then save money by using much less water. The routine spraying of very large amounts of water on landscapes is a recent cultural innovation and only recently became a standard feature of new homes. Too many homebuyers seek a lush lawn and mistakenly believe that excessive watering is what their grass needs.

If a homeowner insists on installing a network of underground irrigation pipes surrounding the home, there are opportunities to at least minimize the amount of water used. Rain sensors that cut off irrigation when there is enough rain are required by state law, but the law is seldom enforced. A major survey of home water use found that only 25 percent of homeowners with in-ground irrigation systems with timers reported that they even had rain sensors.[16] The proportion of properly maintained sensors undoubtedly is even smaller.

Rain sensors are only one of the technological devices that can help save water in landscape irrigation (if they are operational). Other studies have shown the benefits of using soil moisture sensors rather than rain sensors. A soil moisture sensor adjusts the automatic irrigation system to apply less water when the soil has sufficient water for plant roots. It measures not the rain but the actual amount of water available in the root zones of plants, which is where it matters.

However, the sensors are only part of the irrigation system. Even if the sensors do their job, the homeowner's irrigation system may overapply water. In California, for example, a recent study found that many the homes with new "smart" irrigation systems actually increased the amount of water they used because the soil sensor decided that more irrigation was needed than homeowners applied before it was installed.[17] The actions suggested below could save huge amounts of water.

Enforce the Law that Already Requires Operational Sensors on All Automatic Irrigation Systems (Section 373.62, F.S.)

Automatic irrigation systems are water guzzlers and can easily be operated in a way that greatly reduces water use. State law requires automatic irrigation systems to have devices that shut off the system when it is raining. Simply enforcing this law that saves water and money would result in big savings.

The 2007 Drought Smart report proposed one possible measure to improve compliance: a state law prohibiting an irrigation contractor from doing any work to install or repair a landscape irrigation system unless the work included proper operation of a rain or soil moisture sensor. The 2009 legislature considered this recommendation but merely directed local governments to *consider* adopting such an ordinance.

Provide Additional Mobile Irrigation Labs for Landscape Irrigation

This program, already operational in some parts of the state, provides trained specialists to educate homeowners and commercial landscape companies about landscape irrigation efficiency. For the most part, these services are provided free or at low cost. The Drought Smart report concluded that the evaluation done by mobile irrigation labs decreased water consumption at homes by an average of 22 percent. However, the initial savings can decline over time, so repeated visits or other programs may be necessary to preserve initial reductions. The legislature could require water management districts, counties, and cities to provide low-cost services through the mobile irrigation labs. Unfortunately, the trend is the other way: districts and local governments are reducing their expenditures on mobile irrigation labs.

Require Local Governments to Adopt Effective Water-Saving Landscape Ordinances

Model landscape ordinances that save water and protect water quality already exist. For example, the FDEP collaborated with other state agencies, water management districts, local governments, the University of Florida, and other organizations to publish *Landscape Irrigation and Florida-Friendly Design Standards*.[18] The recommendations were published in early 2007, after unanimous approval by a statewide committee. Under current law, however, water users and local governments can ignore the recommendations; the standards apply only to those local governments that decide to have such an ordinance at all. Most do not.

The legislature could simply require all local governments to fully implement the state Landscape Irrigation and Florida-Friendly design standards.

Train, Certify, and License Landscape Irrigation Professionals

It can cost somewhat more to have landscape irrigation systems installed properly but the payoff is in lower water bills and less stress on water resources. Homeowners and other purchasers of landscape irrigation services benefit from receiving a high quality of service. The Water Conservation Initiative and the Drought Smart reports both recommended that the state should license companies and individuals involved in irrigation design, installation, and maintenance. This could be done in cooperation with professional organizations like the Florida Irrigation Society.

Strengthen the Standards for Efficient Use in Landscape Irrigation

As discussed in Chapter 3, the water management districts have sweeping powers to promote efficiency. They should, by themselves or under state leadership, define upward the water use efficiency requirements for landscape irrigation. This would be based on the most basic principle of water use permitting in Florida, that water use permit applicants must demonstrate that their proposed use of water is "reasonable-beneficial":

> "Reasonable-beneficial use" means the use of water in such quantity as is necessary for economic and efficient utilization for a purpose and in a manner which is both reasonable and consistent with the public interest. (Section 373.019(16), F.S.)

A water use permit can be issued only if it is for "efficient utilization" of water and only if the use is "reasonable and consistent with the public interest." Therefore, water conservation is a fundamental precondition for securing a water use permit. The water management districts could decide that additional water conservation measures are necessary to achieve a much higher level of water use efficiency. If a community either does not have an adopted landscape ordinance that requires efficient water use, or does not require licensing of irrigation system installers, the water management district could simply deny the water use permit application because it facilitates the wasteful use of water.

Decide Whether Adult Playgrounds Need 600 Million Gallons a Day

Recreational irrigation for all purposes in Florida uses about 600 mgd of water, half of which is potable. Golf courses are water intensive and are major users of potable water, reclaimed water, and stormwater. Water use at golf courses is about 1,915 gallons per acre per day.[19] Another way of looking at it is that water use is about 2,935 gallons per round of golf.[20] A golf foursome would use close to 12,000 gallons of water for a single round of golf – more water than an average family uses in a month.

Golf is very popular in Florida, which has more courses than any other state. Many golf courses use reclaimed water or collect on-site stormwater for later irrigation. Nonetheless, it must be asked whether the state can afford to apply more than a half-billion gallons of water a day on golf courses and other adult playgrounds.

SAVING WATER INDOORS, TOO

Until recently, the U.S. Environmental Protection Agency has not been a leader in promoting water use efficiency. They have now begun to build up their WaterSense program. This effort is similar to the Department of Energy's very popular "EnergyStar program, which rates the energy efficiency of various appliances and devices. WaterSense actually is much more rigorous in assuring good performance than EnergyStar because WaterSense requires manufacturers to submit the results of third-party independent testing rather than only their own self-interested claims of product performance. Devices that use less water can be certified by the WaterSense program as water-efficient if they also perform at a high level. For example, new high-efficiency toilets that use only 1.28 gallons per flush are certified by third-party performance tests and thereby achieve outstanding levels of customer satisfaction.

Indoor water use per person in single-family homes, with the use of certified WaterSense devices, is expected to decrease, in a water-efficient home, by 21 percent (from 49.9 gpcd to 39.5 gpcd). The WaterSense program recognizes the water–energy nexus (addressed in Chapter 14). Water-efficient single-family homes in Florida are projected to save 3,500 gallons of hot water a year, avoiding the use of 628 kWh of electricity or 3,070 cubic feet of natural gas.[21]

The technology to save great amounts of water and energy is available on the market. Water-efficient devices should become the standard in Florida rather than an unusual option chosen only by some particularly well-informed purchasers.

WATER UTILITIES SHOULD BE LEADING CONSERVATION MANAGERS

Much responsibility for water conservation lies in the hands of end users in public water supply, both indoor and outdoors. However, the responsibility for saving water resides not just at the customer level. The water utility that provides water also has opportunities to promote much higher levels of water use efficiency. Some of those opportunities are described below.

Use the Conserve Florida Guide

The on-line *Water Conservation Planning Guide* at the Conserve Florida Clearinghouse is one of very few in the country specifically designed to meet the needs of a particular region or state. It is intended to provide a menu of practical solutions that utilities, homeowners, and other water users can implement to save water. The Miami-Dade Water and Sewer Department, the state's largest utility, used the *Guide* to develop a 20-year water conservation plan. However, only a very small number of Florida utilities have chosen that path to water efficiency.

The slow pace of acceptance of the *Guide* is similar to the problem with the Landscape Irrigation and Florida-Friendly design standards. The legislature required that a water conservation tool be developed but failed to require that it actually be used. The statute says only that a utility "may propose" a goal-based water conservation plan. The *Guide* is the best available tool to formulate a goal-based plan, but most utilities are deciding not to venture into new territory. The legislature should require that all utilities (except the smallest) develop water conservation plans based on the *Guide*. The first set of conservation plans could come from the larger utilities, with medium-sized utilities to follow.

Require that Inefficient Plumbing Fixtures Be Retrofitted at Time of Home Sale

Simply changing out older water-wasting toilets, faucets, and other devices for modern water-efficient devices can save a great deal of water. The Water Conservation Initiative estimated that swapping new devices for the older ones would save about 40 gallons per household per day. This change-out and device update should be a requirement of property transfer. As sales occur, older facilities would become more water-efficient. Owners would save on water and wastewater bills from the moment of installation.

DeKalb County, Georgia already has such an ordinance. Before home buyers can connect to the city's water service system, they must certify that all plumbing fixtures are water-conserving. That means that either the seller or buyer must arrange for retrofitting immediately before or after closing on the property.[22]

Individual Florida communities can adopt such a requirement, but the legislature would be more appropriate as a policy-making body and should make it a statewide policy. If necessary, local water conservation programs could offer financial assistance, such as toilet rebates, to assist in the upgrades.

Price It Right

The Water Conservation Initiative recommended the use of "conservation rate structures" in utility bills. Conservation rate structures provide a clear price signal to water users: the more water they use, the more they will pay per thousand gallons. Inclining block rates, in which the cost of water goes up in steps as water use increases, are a common type of conservation rate and result in more efficient use of water.[23] Careful design of conservation rates can protect the revenue base of the utility and prevent low-income customers from paying too much, while also promoting efficiency.

One option to achieve effective conservation rates is to authorize the water management districts to order specific conservation rates. Another option is for the legislature to require all utilities in the state to adopt conservation rates, including approval from the appropriate rate-setting authority.

The legislature could go even further and adopt the specific recommendations of the Drought Smart report:

- Define minimum requirements for conservation rates and require their use by utilities in the development of their rate structures. Florida needs real conservation rates, that reward efficiency.
- Phase in conservation rate structures statewide within three years.

AGRICULTURAL IRRIGATION

More water is used in agriculture than in any other water use sector, as is described in Chapter 10, Water Supply Is Easy. Large water savings in agriculture can occur merely by removing the current financial and regulatory incentives for wasteful water use. Additionally, modest water use fees would also strongly encourage more conservation in the agricultural sector. There are many other specific water conservation opportunities in the agricultural sector (with proportionately large potential water savings).

Provide More Effective Mobile Irrigation Lab Services for Agriculture

Mobile irrigation labs were recommended earlier in this chapter to help reduce water use in the landscape irrigation sector. For agriculture, the labs conduct somewhat different evaluations, tailored to the characteristics of particular crop and irrigation systems. Whether a farmer takes advantage of these low-cost services is currently entirely optional. Moreover, a Florida farmer is currently not required by law to implement any of the recommendations from a mobile irrigation lab visit, even if the improvements are cost-effective. The data from the farm evaluation are not even provided to the water management district. This means that the district cannot take advantage of the information in improving current water use practices or in evaluating the next water use permit application.

Rather than leaving the option to a farmer, the state or a water management district should require each large agricultural permittee to have regular inspections from a mobile irrigation lab. The results of each inspection should be transmitted to the relevant water management district. Users who are shown to be inefficient should be required to adopt better water use practices.

Measure It

Agricultural water use would benefit from application of the familiar adage that "If you don't measure it, you can't manage it." Without measurement, a farmer (and the water management district) cannot know how much water is being applied to a crop, whether an expensive leak has occurred, or how much is being spent on irrigation. Some, but not all, of the water management districts require accurate measurement of all agricultural water use. Measurement tends to prevent overwithdrawals of water from limited resources and helps reduce runoff from farm fields. It is a simple and effective tool.

All water use permits above a certain minimum threshold should measure water use accurately and report the results to water management districts. The information should be publicly available and summarized in annual water use reports published by the districts. Only two of the districts (the Southwest District and the St. Johns River District) publish such annual reports at present.

THE POTENTIAL FOR EFFICIENCY

Florida water users and water managers know how to use less water while achieving the same water use benefits. The state could reduce water use dramatically if it implemented recommendations like those described in this chapter.

NOTES

1. "Days of plentiful water have come to an end," *Fort Myers News Press,* 8 November 1985

2. South Florida Water Management District. (2008) *Water Conservation: A Comprehensive Program for South Florida,* p1

3. Tampa Bay Water. (2008) *Compilation of Members' Five Year Conservation Plans: Annual Update, January 31, 2008,* p8

4. City of Oviedo [Florida], "H2O water conservation incentive program," www.cityofoviedo.net/index.php?q=node/1106, accessed 7 April 2010

5. Miami-Dade Water and Sewer Department. (2008) *Miami-Dade Consolidated PWS, Water Use Permit No. 13-00017-W, Compliance Highlights, Report to the South Florida Water Management District, July 7, 2008,* pp1–10

6. Miami-Dade Water and Sewer Department. (2009) *2008 Annual Water Conservation Plan: Conserve Florida Report, March 2009,* pp3-1-2 and 3-1-3

7. San Antonio: Water System. (2008) *Water System 2008 Budget,* p44; Seattle: Public Utilities Department (2008) *Saving Water Partnership 2007 Annual Report, May 2008,* pp1–16; Los Angeles and Las Vegas: William Blomquist (2007) *Water 2010: A 'Near Sighted' Program of Water Resource Management Improvements for the Western United States,* National Water Research Institute, Fountain Valley, CA; Tucson: Tucson City Ordinance 10597, amending Tucson Code Chapter Six, adopted October 14, 2008. Even the best U.S. examples are outdone in other nations. Comparisons should be made carefully because of different socioeconomic and cultural factors. In the United Kingdom, for example, water use is only 40 gpcd. Considering this to be not good enough, they have set a goal of reducing that by cost-effective measures to only 35 gpcd by 2030 (*Future Water: The Government's Water Strategy for England,* February, 2008, Department for Food, Environment, and Rural Affairs (DEFRA), London, p25). In Queensland, Australia, residential use of water was only 34 gpcd in 2008. (*2008 Water Report,* Queensland Water Commission, 2009, City East, Queensland).

8. This list of the benefits of water conservation draws from the August 2008 report of the Conservation Committee of the Florida Section of the American Water Works Association in their Florida 2030 visioning process.

9. Florida Department of Environmental Protection. (2002) *Water Conservation Initiative*, pp1–170

10. Florida Department of Environmental Protection. (2007) *Being Drought Smart*, pp1–33

11. American Water Works Association, Florida Section. (2008) *Our Vision for Statewide Water Supply*, pp1–22

12. Governor's Action Team on Energy and Climate Change. (2008) *Final Report, October 15, 2008*, p8-4

13. Struhs, David B. [FDEP secretary]. (1999) "Memorandum to executive directors of the water management districts, August 3, 1999"

14. Adams, Bruce. (2008) "Water use efficiency is within our vision," *Florida Water Resources Journal*, vol. 60, no. 8, p4. Mr. Adams is the 2010 chair of the Water Use Efficiency Division of the Florida Section of the American Water Works Association.

15. Melissa B. Haley, Michael D. Dukes, and Grady L. Miller. (2007) "Residential irrigation water use in central Florida," *Journal of Irrigation and Drainage Engineering*, September/October, pp427–434

16. Whitcomb, John B. (2005) *Florida Water Rates Evaluation of Single-Family Homes*, prepared for the Southwest Florida, St. Johns River, South Florida, and Northwest Florida Water Management Districts, p37

17. Mayer, Peter, William DeOreo, Matt Hayden, Renee Davis, Erin Caldwell, Tom Miller, and Peter J. Bickel. (2009) *Evaluation of California Weather-Based 'Smart' Irrigation Controller Programs*. Report to the California Department of Water Resources

18. Florida Department of Environmental Protection, Committee on Landscape Irrigation and Florida-Friendly Design Standards. (2010) "Landscape irrigation and Florida-friendly design standards," www.dep.state.fl.us/water/waterpolicy/land_irr.htm, accessed 3 July 2010

19. Obreza, Thomas. (2004) *Florida Springs Land Use Information Tool*, Circular 1448, Soil and Water Science Department, Institute of Food and Agricultural Sciences, University of Florida, Gainesville, FL

20. Haydu, John, and Alan Hodges. (2002) *Economic Dimensions of the Florida Golf Course Industry*, EDIS Document FE 344, Institute of Food and Agricultural Sciences, University of Florida, Gainesville, FL. Haydu and Hodges provide an estimate of golf course water use in 2000 of 172 billion gallons as well as an estimate of rounds played in 2000 of 58.6 million. That equals 2,935 gallons per round.

21. U.S. Environmental Protection Agency. (2008) *WaterSense Program, Water-Efficient Single-Family New Home Specification Supporting Statement*, pp1–7

22. DeKalb County, Georgia Code of Ordinances, Sections 25, 45, 49, February 5, 2008

23. Whitcomb, op. cit., note 16

CHAPTER TWELVE

Charge a Few Nickels for Water

*I*t's free. No one pays a nickel for the privilege of extracting water from an aquifer, river, or lake in Florida. After a water use permit is obtained, the only cost for withdrawal is a pump and the electricity to take the water out of the source. No payment has to be made to the state or water management district for withdrawing 10,000 or even 10 million gallons of water a day. The water is free even if the withdrawal makes water unavailable to other users and even if big profits are made from the water withdrawal. Not charging anything for water leads to the misallocation of water, which is, after all, a valuable economic resource. In contrast to how water is given away, Floridians normally expect to pay for a commodity at a price that bears some relationship to the value they receive. How can a state bring the economic value of water into the decisions made by water users? Two principal alternatives have been considered in Florida: water markets and water use fees.

Markets for commodities have great benefits. The price signal in a market promotes efficiency, which certainly is needed in a state that wastes as much water as Florida. In principle, too, markets tend to move water from uses with relatively low added economic value (like growing corn or hay) to higher-value uses (like cities, power plants, and factories). Water markets, at least in theory, could reduce the damage done by overpumping currently encouraged by the lack of a price for water. Government subsidies for water supply development could be cut back if water were allowed to find its natural economic price among competitive buyers in a market.

With all of these theoretical advantages, why hasn't Florida adopted a system of water markets? The practical water management answer is that the theory of water markets is much different from the complex reality of Florida water management. A full water market, as discussed below, would have crippling disadvantages even if it fostered awareness of the economic value of water. A better way of achieving water use efficiency objectives would be a water use fee. If there were a modest fee for water, Florida's water supply problems would become easier to manage.

THE WATER MARKET ALTERNATIVE?

"I don't want any company selling me something that comes from God."[1]

State senator Skip Campbell's words spoke for many when he opposed an effort by Governor Jeb Bush to create a water market in Florida. In 1999 Azurix, a subsidiary of the Enron energy trading company, tried to persuade Florida politicians to set up a water market.[2] The company employed lobbyists and a former executive director of the South Florida Water Management District to help the proposal along and met with Governor Bush in his office. The governor's meeting with the lobbyists occurred just before Enron's fraudulent bookkeeping was revealed and the company proceeded to financial bankruptcy and the imprisonment of company executives. Azurix's idea was that both the company and the state would make money off of Everglades restoration − if only the company were granted the exclusive right to sell water from the restoration projects that Azurix would build. The Everglades water market proposal never progressed very far due to political opposition and the financial disintegration of both Enron and Azurix.

The Azurix misadventure was only the most gaudy of recent efforts to graft a water market onto the fundamentally regulatory approach of Florida water management. The 1971 Model Water Code, which was the basis for the Florida Water Resources Act of 1972, cautioned that any property right in water was severely limited by the natural fluctuations in the availability of water and the competing claims of other users. Nonetheless, in the early 1990s resource limits in the Floridan Aquifer southeast of Tampa spurred a number of academic proposals to create a system for the "exchange of water rights" in that region. In 1997, the legislature's Office of Program Policy Analysis and Governmental Accountability recommended that the state consider "voluntary reallocation" because it "allows markets to help determine what the most economically efficient uses are at any given time, while still protecting the environment and existing legal users."[3]

The 2002 state Water Conservation Initiative tried to balance both sides of the controversy by recommending the "use of market principles" while "still protecting the fundamental principles of Florida water law."[4] Under Governor Bush's direction, the secretary of the Florida Department of Environmental Protection continued to try to find a way to structure sales of water use allocations. The idea was to sell, or auction off, blocks of water in long-term permits to the highest bidder. The purchased water rights would be held and owned as a kind of long-term lease rather than a straight-out permanent sale of water rights.[5]

It turns out, however, that water has many special characteristics that make it difficult for a market in water to function well. Water, as a substance and as a Florida natural resource, is different from other economic commodities. Uniquely, it is at the same time a natural resource, a product in commerce, a carrier of waste products, and an essential element for ecosystems and human life. Formal markets for water seem like good ideas on first examination but there are complications. These multiple obstructions are especially troublesome in Florida due to its unusual hydrology and history of water management practices.

Water is Symbolic

Water is a primary symbol of life. In different contexts, it symbolizes continuity or the connection between all beings. Water is used in the religious ceremonies of many cultures. Lake Okeechobee is the metaphorical "liquid heart" of south Florida. Sixteen state parks are named after springs. The principal aquifer on which much of the state depends is named after the state itself. When freighted with all of the symbols that water carries, it is not regarded by most as a simple commodity like any other, to be bought and sold freely on the market.

Water is both a Public Resource and a Private Good

Water moves easily between being a resource of the public at large and a privately owned good. For example, water is a public resource in Lake Okeechobee. No one owns the lake and all have an equal opportunity to benefit from this common resource. A person bass fishing on the lake is using the same public resource as a city or a farm drawing drinking water out of it. The resource is available to all.

On the other hand, when water is withdrawn from Lake Okeechobee and supplied to a sugarcane farm in the Everglades Agricultural Area, the water is instantly transformed into a private good. The sugar farmer seeks to get as much private benefit out of the water delivered to the farm as possible and is not personally responsible for the public good aspect of the water remaining in the lake. Some of the irrigation water ends up going back into the lake, at which time it is transformed back into a public good. The same water sometimes is more of a public good than a private good, and sometimes more of a private good than a public good. Designing a water market faces difficult challenges when confronting these dual and interchanging qualities.

Water Moves Around

Rain that falls on the springshed of the Silver River in Marion County may land on a nursery business and meet the water needs of plants. A fraction of the rain, however, penetrates the soil surface and becomes part of the Floridan Aquifer. Years (or even decades) later, it emerges in the vent of Silver Springs. The Silver River joins the flow of the Oklawaha River, is impounded for a time in Rodman Pool, flows farther to join the St. Johns River, and proceeds slowly downstream to the estuary near Jacksonville. Along the way, human activities sometimes remove water from the river system and sometimes add water. Downstream interests, like Jacksonville at the mouth of the river, are very interested in protecting upstream flows because diverting water from the river upstream could harm the estuary. The different political jurisdictions in the watershed are all interested in having a voice in water management, which might be stifled if the primary medium of communication were a market.

How can a water market allow all of the interests that have a stake in water, upstream and downstream, to participate in pricing this mobile resource?

Water is Cheap but Heavy

Modern societies use enormous quantities of water. Floridians use more than a thousand gallons of water a day (combining freshwater and saltwater). Looking only at the public water supply sector, an average of 158 gallons a day per capita is treated to drinking water quality in Florida cities. That's close to a ton of drinking water delivered daily per person for less than a dollar.

Large quantities of water are available almost everywhere in Florida, even though the water sometimes contains salts that must be removed to make it drinkable. It often is cheaper to treat low-quality water to make it potable than to pay the energy cost of long-distance transport. A water market may fail before it starts simply because water is inexpensive to obtain but relatively expensive to transport.

Transaction Costs can be High

It could be very difficult to complete the deal in a real water market rather than a hypothetical one. The buyer must expend effort to determine that water is actually available from the seller in the claimed amounts and at the appropriate quality. Parties that may be affected by the transaction could file legal challenges about possible environmental effects and slow up the process, thereby adding uncertainty and costs. Even in a system that facilitates the purchase of water rights, some governmental entity would have to be involved to safeguard against damaging levels of withdrawals. Only government oversight can address the social and environmental benefits of water, which are not reflected adequately in the market price of water.

A property rights-based market regime may or may not be simpler to run than a water use permitting system. It depends on the transaction costs and how much regulatory oversight remains, even with a market.

A Gallon of Water is not a Gallon of Water

Suppose that the owner of an orange grove in a rural area had a property right to pump 10 million gallons of water a day from a well. And suppose the farmer sold that withdrawal right to a city 20 miles away via a water market system. That apparently simple transaction is in fact very complex because the withdrawal of a gallon of water in one place is not equivalent to withdrawing a gallon in another location. What are some of the complicating questions that have to be addressed?

• Does it matter if the city's withdrawal is from a different aquifer than that of the farmer?

- Does it matter if the city is withdrawing the water in a more mineralized portion of the same aquifer as the farmer? Or closer to the shoreline and tending more to induce saltwater intrusion from the ocean?
- If the city's point of withdrawal tends to move a contaminant plume toward it from a waste site, how should that affect the transfer?
- Would the city's pump dry up a nearby spring?
- Would the city return more or less water to the aquifer after using the water than the farmer?
- Would there be more or less pollution from the city's stormwater runoff and groundwater recharge than from the farm operation?
- Would net economic activity in the region be larger or smaller after the transfer?
- Would either activity be more in the public interest?
- Does it matter that the city has access to slightly more costly reclaimed wastewater but would prefer to avoid using that source?

One gallon of water is different from other gallons of water because water is so closely bound to other natural resource functions and to many human uses.

Water Use Permits are Very Specific

In the example above, the farmer is negotiating to sell the rights in an agricultural water use permit to a city. It may turn out, however, that the commodity being purchased – the water use permit – is much more complicated than the city might imagine. On going to the city council for final approval of the purchase, the water utility director might face some tough questions, such as:

- The farmer's water use permit has a limit on 5-year average annual withdrawals and a separate higher limit on drought year withdrawals. That made sense for the farmer because the crops needed more irrigation water during droughts. Do we get to keep that higher allocation? How do we use it?
- What is the city, as the new holder of the permit, to do with the extra daily amount allocated to the farmer for occasional frost protection during rare winter freezes?
- How do we know that we won't violate the minimum flow and level for water bodies and aquifers in our own region?
- The farmer has a permit to withdraw 10 million gallons a day but never used more than 3 million gallons a day. Are we buying the right to withdraw 10 million or 3 million gallons a day?

Water permits are not fungible commodities like sugar or corn or aluminum. Often, a valid water use permit is so closely tailored to the needs of the water user and the associated water resources that it is very difficult to exchange an allocation with the water needs of another user at a different location (or even at the same location). ASCE's guide to implementing their Model Riparian Water Code notes correctly that it can be challenging to "define exactly what is being transferred."[6]

Using the Wrong Mental Models

For water markets to work at all, they must make sense from a hydrological standpoint and be adapted to the water realities of the region. A water market designed for Florida must not be based on assumptions more suitable for Arizona. For example, one might imagine how a large reservoir of water in a western state, built solely for water supply, could have the contents auctioned off in a market. The reservoir is a known stock of water, of known quality, constructed to be put to use for economic activities. Why not auction it off to the highest bidder – the one that is willing to vote with actual money to demonstrate the value of the water?

This mental model makes little sense in Florida. There are very few large surface-water reservoirs. Almost all of them were built by a single unit of government for their own utility's use. There is no set of reservoirs ready for the state to declare that bidding on them is now open.

Sometimes a particular aquifer is imagined as the place in which a market could operate. That oversimplifies or misunderstands Florida hydrology. Most aquifers in Florida are not like a "common pool" resource with few connections to other resources or undesirable externalities. Groundwater in Florida usually is connected closely to surface features like wetlands and lakes. Removing too much water from an aquifer drops the bottom out of the surface water resources. As noted in Chapter 1, The Watery Foundation, Florida surface water and groundwater are closely connected and interchange all the time. A water market is an idea, whereas water resources are physical realities. The idea does not match very well with the specific character of Florida lakes, rivers, and aquifers.

Real Water Markets, Under Governmental Oversight?

If a water market were to exist in Florida, it would rely upon substantial government involvement. Government would be necessary for creating legal rights for buyers and sellers, for overseeing disputes about legal claims, for preventing fraud, for deciding on the levels of allowable water withdrawals, and for taking into account the important environmental and social effects of withdrawals. In effect, a water market is possible only under strong governmental oversight.[7] It is not a choice between onerous governmental oversight and an efficient water market operating in the private sector.

The Overselling of Water Markets

As appealing as water markets may be in theory, they are remarkably rare in practice. Water markets are usually promoted as the next great thing, just over the horizon, but they seldom arrive anywhere. Despite theoretical benefits, genuine functioning water markets are rare. They hardly exist in any state or country. Joseph Dellapenna's and Steven Draper's review of the American experience found that, despite many academic articles advocating water markets, they are rare in practice: " . . . true markets, without heavy-handed state intervention to determine who can buy or sell, from or to whom, at what price and for what purposes – have always been rare and small-scale for bulk water."[8] Dellapenna asked a fair question, "If markets for water are so good, why are they so seldom used?"[9]

Carl Bauer's studies of the free-market water law of Chile provided some answers. That national water market system, he notes, "is so pure a symbol of free-market theory and ideology that both proponents and critics have had a lot of stake in whether the resulting markets are considered a success."[10] Among his conclusions, after a decade of study, was that market incentives have not worked as expected and that significant market activity was limited to only a few areas. In Bauer's view, "The Chilean model has gone too far in the direction of unfettered regulation" and hasn't thought through the public interest.[11]

One reason why water markets continue to be proposed is that imagined water markets, in their perfection, look better than the complex and conflictual reality of water management. A kind of reverse Murphy's Law is often assumed to operate for water markets: if anything could go wrong, it won't. In the real world, water markets would go wrong much more often than in theoretical perfection.

THE WATER USE FEE ALTERNATIVE

The 1971 Model Water Code noted the possibility of a water use fee and that:

> If the state were to exact periodic fees for the use of the water, it is conceivable that such a schedule of charges could be established which could discourage wasteful amounts of use and perhaps even some uses at wasteful locations or for wasteful purposes. There appears to be no doubt that fees may constitutionally be charged by the state in connection with new uses.[12]

Although the Model Water Code did not recommend a direct water use fee, the authors did recommend what they called an "annual surveillance fee" to be imposed on each holder of a permit and which would assist in water resource monitoring. Unfortunately, no "surveillance fee" was included in the Florida Water Resources Act of 1972. The water management districts charge a fee for the initial permit application but nothing at all after the permit is issued.

The idea of a water use fee is powerful enough that it is a central feature of the *Regulated Riparian Model Water Code* adopted by ASCE in 2004. Intended as a model for all states like Florida that base their water law on the Eastern doctrine of water law, the ASCE Model Code is very direct about the need to implement a water use fee:

> Without requiring fees for the value of the water used, one cannot really hope to achieve real efficiency in the use of water and therefore of ensuring sustainable development[. . . .] In any event, it is important that water users cease to consider water a "free good" if we are to achieve a high measure of efficiency in the use of water and to see the application of water to its higher valued uses.[13]

Although Florida does not have a system of water use fees today, the idea has been seriously considered a number of times. Frank Caldwell, the first chair of the governing board of the Northwest Florida Water Management District, regretted that user fees were not the selected funding mechanism. He pointed out that it would have cost only "pennies per thousand gallons" and would have been more equitable than the property tax levy.[14]

In December 1976, in the same year that Florida voters authorized the water management districts to levy a property tax, the executive director of the Southwest Florida Water Management District suggested that any expansion of pumping from the Starkey wellfield in Pasco County should be accompanied by a fee of 6 cents per 1,000 gallons. The executive director proposed that fee revenues would be used by the Southwest Florida District as "seed money" for water projects and additional wellfield studies.[15]

A governing board committee of the Northwest Florida Water Management District recommended in 1983 a water use surcharge as an alternative to property tax funding.[16] (A water use fee is especially meaningful for the Northwest District because the constitutional amendment authorizing property taxation by the districts limits the Northwest District a cap only one-twentieth as high as the other four districts.)

A major milestone in considering a water use fee came when Governor Bob Martinez's Water Resource Commission recommended a water use fee in 1989. They evaluated, among other options, an "equalization fee" that imposed higher charges on those who used more water than average.[17] The commission laid out in some detail the purposes to which the revenues would be applied:

> Recommendation 18: Collect a fee from all users based on water used. Credits shall be given for aquifer recharge, use of reclaimed water, reverse osmosis, desalination, or other alternative technologies. Funds shall be accrued in a Water Resource Trust Fund to be used for the following purposes:

- Alternative sources development (reverse osmosis, reclamation, conservation, etc.) within critical water supply problem areas.
- Promoting of area-wide water supply authorities and reuse systems through planning studies, start-up funding, or low-interest loans.
- Resource protection activities, such as wellhead protection and recharge area protection. Priority for funding should be given first to resource protection activities in "donor" water supply areas and second to the recipient critical water supply problem areas.
- Water quality testing mandated for public water supply systems.
- Infrastructure improvement or regionalization.
- Incentives for conservation for all users.

The Governor's Water Resource Commission also considered an "equalization fee" for those who used more water than average. This would exempt from any fee those water users who use less than the average amount of water. The effect would be to encourage more water conservation in the high user class.[18] Detailed evaluations of the commission's ideas were conducted by 1991 by Chase Securities. They studied fee scenarios of 2 cents, 10 cents, and 20 cents per 1,000 gallons.

Chase concluded that a fee on the public water supply and the agricultural and industrial sectors was feasible and would have minimal adverse economic effects.[19] In 1990, the state Department of Environmental Regulation evaluated a fee of 1 cent to 4 cents per 1,000 gallons on public water supply to fund its operations.[20]

Governor Lawton Chiles, the successor to Governor Martinez, also gave a water use fee serious consideration. During his term a bill on water use fees was introduced in the legislature but it did not succeed.[21] The South Florida Water Management District proposed a water use "surcharge" for both residential and agricultural users.[22]

In 1993, the legislatively established Partners for a Better Florida Advisory Council followed up and recommended that water use fees be established and credited to a water use and conservation trust fund. In 1997, the Florida legislature's Office of Program Policy Analysis and Government Accountability noted that an advantage of a water use fee would be that it would "help to infuse a greater consideration of the value of the resource into the water supply decision-making process."[23] Among the disadvantages they listed was "Fees could produce somewhat less revenues than may be expected because the fee would likely cause some reduction in use." (This "disadvantage" is the converse side of an advantage. A reduction in water use would mean only that water users were in fact responding to the price signal about the economic value of water, as the water use fee is intended to accomplish.)

In 2008, the idea of a user fee was again considered, this time by a committee in a public water supply industry "visioning process for 2030" in Florida. The Conservation Committee for the Florida section of the American Water Works Association firmly recommended a water use fee:

> Recommendation: Florida should impose fees on all water users, as recommended by the 1989 Governor's Water Resource Study Commission and the American Society of Civil Engineers. The price signal will help promote increased water use efficiency. As in other economic sectors, a price on water related to use will help make water available to other users by discouraging waste. The fee could be used for programs to protect water resources, develop new alternative supplies, or fund water use efficiency implementation.[24]

Despite all of these various recommendations, no water use fee has yet been established.

OTHER USE FEES IN FLORIDA

Although Florida does not have a system of water use fees, it has not hesitated to impose use fees for other natural resources. For example, miners pay for the privilege of extracting valuable minerals. Florida levies "an excise tax upon every person engaging in the business of severing phosphate rock from the soils or waters of this state for commercial use." (Section 211.3103, F.S.) The tax of $3.325 per ton is dedicated to multiple purposes, including local governments, phosphate mining research, mine reclamation, and public land acquisition. Titanium miners (operating in northeast Florida) must make a payment of $3.11 to the state of Florida for every ton of ore they extract. Another mining example is for limerock and sand mining in the "lake belt" in western Miami-Dade County. A fee of 24 cents a ton is levied on each ton of limerock or sand mined. An additional fee of 15 cents per ton is levied until enough revenue has been collected to pay for a water treatment plant for water from the Northwest Wellfield county. (Section 373.41492, F.S.)

Southwest and far northwest Florida have significant oilfields for which royalties are paid to the state. The two oilfields pay an 8 percent royalty per barrel extracted from the ground. An additional "pollutant tax" is levied on the production or importation of all petroleum products (as well as some other products):

- Coastal Protection: 2 cents per barrel of petroleum
- Water Quality: 5 cents per barrel of petroleum products, pesticides, and chlorine
- Inland Protection: 80 cents per barrel
- Hazardous Waste Management: 5 dollars per gallon of perchlorethylene[25]

The state of Florida also leases the use of underwater land for various purposes such as marinas. The basic price for leases is 6 percent of the annual income derived from private use of state property, with higher rates for activities in state aquatic preserves. A party wishing to dredge state waters has to pay at least $1.25 for every cubic yard dredged, and up to $3.25 in some designated counties.[26]

The state and local governments also impose other resource use fees. For example, a stormwater utility is a mechanism to fund the cost of municipal stormwater management services. It operates like an electric or water utility. The fee is often proportioned to the area of impervious property for larger parcels but may be a flat fee for residences. Stormwater fees for residences in Florida vary widely from less than a dollar a month to several dollars, with commercial properties paying substantially more.

For more than a century, Floridians have been taxing themselves to remove excess water from farms and cities. Under Chapter 298, Florida Statutes, local water control districts (formerly called drainage districts) can build canals, pumping stations, and other works to benefit landowners in the district. The districts are granted the power to levy ad valorem property taxes for this purpose. The amount of taxes varies among the districts, and within districts, based on the benefit to particular property owners. In effect, this is a tax or fee designed by the water control district board to be proportional to the flood control benefit within district boundaries.[27]

Lastly, the Everglades Forever Act imposes a "tax" on agriculture in the Everglades Agricultural Area to limit phosphate pollution, but the Internal Revenue Service concluded that the so-called Agricultural Privilege Tax was really a fee rather than a tax. The IRS reasoned that the charges were not levied at the same rate on all properties and property owners could reduce their charge by demonstrating that the phosphorus runoff from their property was minimized.

The IRS also noted that the statute requires that the charge be different for different areas "based upon a reasonable relationship to benefits received."[28] In FY 2009, the tax raised $10.8 million. The revenues are less than they might have been because farmers had reduced their phosphorus discharges. In this case, the Agricultural Privilege Tax is working in the same way that a water use fee could work: providing a price signal that causes desirable behavior. For the EAA, farmers pay less if they pollute less. With a water use fee, the same incentive would apply.

WATER USE FEES OUTSIDE FLORIDA

Thus far, the water use fee proposal for Florida has not gained adequate traction. Other states and countries have gone forward with water use fee systems:

New Jersey, for example, requires a water use fee of a penny per thousand gallons for water delivered to consumers by public water suppliers. The funds, amounting to millions of dollars a year, are placed in the state Safe Drinking Water Fund. (New Jersey Administrative Code 58:12A-21) *Kansas* has a similar program but assesses three cents per thousand gallons as a Water Protection Fee on public water supply systems, industrial uses, and stock watering. (Kansas Statutes, Annotated, 82a-954, 2004) An additional Clean Water Protection Fee of three cents per thousand gallons is assessed on public water supply systems.[29] The 2008 session of the *Iowa* legislature directed that water use permit fees be charged to cover the reasonable cost of reviewing applications, issuing permits, and ensuring compliance. However, the total amount collected is not to exceed $500,000.[30] *Minnesota* charges a fee based on actual use of water. Above 500 million gallons per year (1.37 million gallons per day average), the fee is $8.00 per million gallons with a maximum of $750 for any agricultural permit. Interestingly, this price signal to promote efficiency even has a sophisticated seasonal adjustment:

> A surcharge of $30 per million gallons will be applied to the volume of water used in each of the months of June, July, and August that exceeds the amount used in January of each year. The summer surcharge applies to municipal water use, irrigation of golf courses and landscape irrigation. This is a surcharge in addition to the regular fee rate based on the yearly total volume used.[31]

In *Texas,* the Barton Springs/Edwards Aquifer Conservation District assesses water use fees based on permitted volumes of water:

- $0.17 per 1,000 gallons for wells or aggregate of wells
- $1.00 per acre-foot for agricultural wells

An additional special Texas law directs the Aquifer Conservation District to assess a water use fee on the City of Austin. For fiscal year 2007, the fee generated $717,320 based on pumpage of 2.8 million gallons of water. A separate annual fee of $0.31 per thousand gallons is assessed for water permitted to be transported outside district boundaries.[32] Other proposals for water use fees in Texas have been considered. Researchers at Texas Tech University concluded that water use fees could conserve water more efficiently than aquifer withdrawal caps.[33] It is also interesting that the Texas Controller of Public Accounts proposed in 2008 that the state consider a general water rights fee of $1.50 per acre-foot of water.[34]

The *Kentucky* River Authority also charges a fee for water use. The Authority has the responsibility to develop a "unified long-range water resource plan" for the entire watershed of the Kentucky River. To assist in carrying out their responsibilities, the Authority may "collect water use fees from all facilities using water from the Kentucky River Basin, except those facilities using water primarily for agricultural purposes." (420 K.A.R. 1.040) All water withdrawn is charged a base fee of 2.2 cents per thousand gallons. An additional fee of 6 cents per thousand gallons is charged for withdrawals from the main channel of the Kentucky River and is dedicated to operate and maintain the locks and dams on the river.[35]

In *California*, the state Water Code directs each person with a "permit or license to appropriate water" to pay an annual fee to the State Water Resources Control Board. The fees are directed by statute to be sufficient to cover the entire costs of the permit program, including compliance activities, planning, and monitoring. (Section 1525 of Code) In 2010, the base annual fee is $100 per permit plus 2.3 cents per acre-foot greater than 10 acre-feet.[36] The fees are periodically adjusted and deposited in the state treasury.[37]

Another California example of relevance to Florida is the opening of a water bottling plant in Lake County, Florida in 2008. The Niagara Bottling Company built a new Florida facility to expand their operations nationally from their California base. In California, the Niagara Company paid more than $320,000 in 2007 to the Chino Basin Water District for the use of 1,106 acre-feet of water.[38] That amounted to about 90 cents per thousand gallons of raw water supplied to the bottler. It is hard to see why a bottling company should pay for each gallon of water in California but not in Florida.

Thus, water use fees have been put in place in many locations in the United States. They could make a difference in Florida, particularly if they are well designed to promote economic and water use efficiency.

A FEASIBLE FEE STRUCTURE

In Florida, water use fees have been considered a number of times but never quite came to be established. Under the state's water use permitting system, a fee would be simple to administer. In fact, it would be much simpler to manage in Florida than in most other states. All water use permittees are registered with the water management district. A fee could be submitted easily with the already required monitoring reports. The fees could be structured in several ways:

1. *A fee on permitted withdrawal capacity.* This would be the simplest fee to administer and calculate. If a permittee had authorization to withdraw a million gallons a day, the water use fee would be based on that capacity, regardless of how much water was actually withdrawn. From a standpoint of economic efficiency, this

assessment is inferior to a fee on actual water use because capacity is being taxed rather than actual use. On the other hand, the capacity to withdraw water does have a relationship to withdrawals, particularly when there is a price put on capacity. Additionally, a fee on capacity would tend to prevent parties from applying for water use permits they do not intend to use and then "bank" water that could otherwise be made available to other parties.

2. *Different fees on different types of water use.* Fees could be higher or lower on withdrawals for different classes of use. For example, the fee could be lower on a user who returned a higher fraction of withdrawals to the source of withdrawal. Another adjustment could be for the water quality effects associated with the withdrawal. In practice, however, assessing different fees on different classes of users could be difficult to calculate and manage.

3. *Different fees for different regions or hydrologic circumstances.* The different water management districts could choose different fees to reflect their unique combination of water resource management needs and financial circumstances. A variation of this approach might be to match the fee to the seriousness of a drought. When water is in shortest supply and efficiency is most needed, the fee could be increased as a "drought surcharge" (like that described above for Minnesota). This could have the effect of significantly reducing water use in a drought.

4. *Volumetric fees for end users.* The more water used, the higher the water bill. This price structure is common in public water supply systems. It provides a direct price signal to users and makes the fee directly proportionate to the amount of water removed and put to use. Utility customers are well aware of this billing system and an additional fee on the simple volume of water withdrawn would be no surprise.

A 2005 report to the Canadian Council of Ministers of the Environment reviewed these four fee options as well as other possible structures. On balance, simplicity in a fee structure was recommended over more complex structures. Simple rate structures like volumetric fees did not have the theoretical economic advantages of more sophisticated systems but had almost as much benefit.[39]

University of California–Berkeley researchers reached similar conclusions in 2010 about a possible "public goods charge" for water in that state. They recommended a volumetric fee on water use in California that would raise $680 million a year. The fee would create a price signal to encourage water conservation and provide a stable funding source for energy and water conservation programs.[40]

Florida's five water management districts collectively raise in the range of $1 billion a year in property taxes. A water use fee could offset some or all of that tax burden. If a user fee makes economic sense, is relatively easy to administer, and could offset an unpopular property tax, why has it not become law? One possible explanation may be anxiety about any novel fee proposal. However, as noted above, Florida already assesses use fees, or "severance" taxes, on a variety of natural resources.

Probably an even bigger obstacle is connected inversely to the potential benefit of a water use fee: the biggest water users would encounter a price signal for excessive withdrawals. The relatively small number of large water users are cities, large farms, and big corporations. They could bring their political power to bear in opposing a fee. For example, the Florida Farm Bureau, representing the single largest user of water, has adopted a position formally opposing any water use fees for their withdrawals of water.[41]

In another example, the Florida Department of Environmental Protection proposed that the 2009 legislature impose a 6 cent per gallon fee on water bottling (much less than the 7 percent sales tax that the state of Washington imposed on bottled water in June, 2010[42]). The fee would not be on the water in the bottles but on the total amount of water used in production, and therefore akin to traditional Florida "severance" fees on oil and phosphate ore. The intended revenues would build up the Water Protection and Sustainability Trust Fund to the original $100 million–a-year levels in 2006.[43] The most potent opposition to a water use fee for bottlers did not come primarily from the bottlers, but from agriculture and other users who feared that a precedent set by any water use fee on any water use could eventually be applied to them.

Although the 6-cent-a-gallon fee (equivalent to $6.00 per thousand gallons) on bottled water alone would raise $100 million, a much smaller fee per gallon would raise the same amount of money if it were applied to larger categories of water use. For example, a fee of only 4 cents per thousand gallons would raise $100 million if it applied to all freshwater use. If private domestic wells were exempt from the policy, the fee would have to rise to only 4.1 cents per thousand gallons for the same revenue level. If the fee were restricted solely to public water supply systems, a fee of 10.8 cents per thousand gallons also would raise $100 million.[44]

USING THE WATER USE FEE

A water use fee makes sense because it is broad-based and provides at least a small incentive for water use efficiency. Large users, especially those with big water conservation opportunities, would have a greater incentive to conserve. Nonetheless, a water use fee should not become the only source of revenue for water management districts. Water supply is only one of the many missions of the districts. A water use fee is not appropriate as the sole source of revenue for other critical water management functions like floodplain management and water quality protection.

What could the revenues from a water use fee be used for? The water use fee revenues could be used directly to reduce property taxes paid to water management districts. Under the current statutory arrangements, one business may use 100,000 gallons of water a day while another may use 10 million gallons a day. Nonetheless, the property taxes the two businesses pay to the water management district may be identical because the taxes are proportionate to assessed property value rather than the amount of water they use.

Another possible application of the water use fee revenues could be to dedicate them to improve water use efficiency, either in particular regions or statewide.

Water conservation could be advanced significantly if it were funded by "conservation fees" on water use, just like energy conservation is funded on electric bills. A fee of a few nickels per thousand gallons would not be a major increase in the cost of water, particularly if there were a compensating decrease in property taxes. However, it would discourage wasteful use of water and demonstrate to Floridians the connection between water use and the cost of managing that resource. The state should seriously consider implementing a water use fee.

NOTES

1. "Texas company wants to help restore the Everglades," *Orlando Sentinel,* 12 November 1999

2. Perin, Monica. (1999) "Azurix makes waves in Florida water fight," *Houston Business Journal,* December 3, 1999. The head of Azurix eventually had to sell off the company assets following the Enron debacle and is now the CEO of the Water Standard Company. This firm tried to interest Florida and other coastal cities in buying water from company ships anchored offshore that would pipe onshore a supply of desalinated seawater. See "Water standard: A global water solution – In partnership with nature," www.waterstandard.com/index.htm, accessed 17 December 2009

3. Model Water Code, pp166–167; Saarinen, Phyllis, and Gary D. Lynne. (1993) *Allocating Water under Scarcity Conditions in Florida: Evaluation and Prospectus,* Economics Report ER 93-2, Food and Resource Economics Department, University of Florida, p ix; Office of Program Policy Analysis and Governmental Accountability, Florida Legislature (1997), *Review of the Economic Components of State Water Policy,* Report 96-82, p8

4. Florida Department of Environmental Protection. (2002) *Water Conservation Initiative,* pp1–170

5. Trigaux, Robert. (2005) "Corporate strife touched Florida," *St. Petersburg Times,* 14 January

6. Eheart, J. Wayland. (2002) *Riparian Water Regulations: Guidelines for Withdrawal Limitations and Permit Trading,* ASCE Press, Reston, VA, p43

7. Bell, Stephen, and John Quiggin. (2006) *The Metagovernance of Markets: The Politics of Water Management in Australia,* Murray Darling Program Working Paper: M06-6, Schools of Economics and Political Science, University of Queensland, Brisbane, Australia

8. Dellapenna, Joseph W., and Stephen E. Draper. (2003) *Straight Talk about Markets for Water,* prepared for inclusion in the Georgia legislature's Joint Comprehensive State Water Plan Joint Study Committee archives, August 2003

9. Dellapenna, Joseph W. (2000) "The importance of getting names right: The myth of markets for water," *William and Mary Environmental Law and Policy Review,* vol. 25, pp317–377. Dellapenna drew the same conclusion about the paucity of operating water markets in a later paper, "Climate disruption, the Washington consensus, and water law reform," *Temple Law Review,* vol. 81, no. 2, pp383–432

10. Bauer, Carl (2004) *Siren Song: Chilean Water Law as a Model for International Reform,* RFF Press, Washington, DC, p.74

11. Barrionuevo, Alexie (2009) "Chilean town withers in free market for water," *New York Times,* 14 March

12. Model Water Code, p122

13. American Society of Civil Engineers (ASCE). (2004) *Regulated Riparian Model Water Code*, Section 4R-1-08, ASCE/EWRI 40-003, ASCE, Reston, VA, pp48–49

14. Oral History Program. (2004) Frank Caldwell, 7 December

15. Snyder, David M. (1976) "Are pay-by-the-gallon water bills in the future?" *St. Petersburg Times*, 2 December, p14B

16. McCartney, Bill. (1991) *A Water Management Year in Review*, 10th Water Management Conference Summary, Northwest Florida Water Management District, Havana, Florida

17. Garner, James III [Chair, Governor's Water Resource Commission]. (1989) Memorandum to Governor's Water Resource Commission, Summary of September 11, 1989 Meeting, September 27, 1989

18. Garner, ibid.

19. Chase Securities. (1991) *Capitalizing a Water Resources Trust Fund with Water Use Fee Revenues: Feasibility and Effects*, JP Morgan Chase, New York

20. Armstrong, Randy. (1990) Memorandum to Dale Twachtmann, September 10, 1990

21. Willon, Phil. (1991) "Paying for the future," *Tampa Tribune*, 11 May. " 'The Governor is considering a state water-consumption fee on homeowners, industry and farmers,' Lt. Gov. Buddy McKay said Friday[. . . .] The proposed fee, a tax the state would collect based on the number of gallons used, would pay for expensive procedures to make salt water drinkable and encourage conservation."

22. U.S. Water News. (1990) "S. Florida district proposes surcharge," http://dloc.com/WL00002378/00001, accessed 14 January 2010

23. Office of Program Policy Analysis and Government Accountability, Florida Legislature. (1997). *Review of the Economic Components of State Water Policy*, Report No. 96-82

24. The author was on this committee and recommended a water use fee.

25. Florida Department of Revenue. (2010) "Florida's pollutant tax," http://dor.myflorida.com/dor/taxes/pollutant.html, accessed 11 July 2010

26. Committee on Environmental Preservation and Conservation, Florida Senate. (2008) *Submerged Land Leasing*, Interim Report 2009-112

27. Anderson & Carr, Inc.(2008) *Appraisal of Real Estate and Other Assets of the United States Sugar Corporation*, for the South Florida Water Management District, November 1, 2008, Anderson & Carr, Inc., West Palm Beach, FL

28. U.S. Department of the Treasury, Internal Revenue Service. (2000) Memorandum from Deborah A. Butler, April 7, 2000, Subject TL-N-5122-99-WLI2.wpd, pp1–17

29. Kansas Department of Revenue. (2010) "Notice 04-08, Clean Drinking Water Fee – Election Available," www.ksrevenue.org/taxnotices/TN0408.htm, accessed 10 July 2010

30. Iowa General Assembly, 2008, House File 2672

31. Minnesota Department of Natural Resources, Water Appropriations Permit Program. (2010) "Water use fee rates," www.dnr.state.mn.us/waters/watermgmt_section/appropriations/feerates.html, accessed 1 July 2010

32. Barton Springs/Edwards Aquifer Conservation District. (2008) "Fiscal year 2009 fee schedule, effective September 1, 2008," www.bseacd.org/, accessed 20 March 2010

33. Johnson, J. W., P. Johnson, K. Rainwater, E. Segarra, and D. Willis. (2004) *Evaluation of Water Policy Alternatives Intertemporal Allocation of Groundwater in the Southern High Plains of Texas*, 2004 Conference of the Universities Council on Water Resources, Portland, OR, July 20, 2004

34. Texas Controller of Public Accounts. (2009) *Liquid Assets: The State of Texas' Water Resources*

35. Kentucky River Authority. (2010) "Water use fees," http://finance.ky.gov/NR/exeres/64F2BA3A-AC9C-4C44-879D-EAE7611DDA9A,frameless.htm?NRMODE=Published, accessed 11 July 2010

36. California Division of Water Rights. (2010) "Fiscal year 2009–10 fee schedule summary," www.waterboards.ca.gov/waterrights/water_issues/programs/fees/docs/fee_schedule_fy0910.pdf, accessed 11 July 2010

37. California State Water Resources Control Board. (2007) "Water rights and water quality certification fees program FAQs – FY 2007–2008," in "Questions related to the Board of Equalizations (BOE) Notice of Determination (Water Right Fee Bill)," www.waterboards.ca.gov/waterrights/water_issues/programs/fees/faq.shtml, accessed 11 July 2010. The California Supreme Court affirmed the constitutionality of this fee. California Farm Bureau Federation v. State Water Resources Control Board (Case No. S150518), Januay 31, 2011

38. Chino Basin Watermaster. (2007) *Final Assessment Package, Pool 3 Assessment Fee Summary*, Chino Basin Watermaster, Rancho Cucamonga, CA, p1

39. Sawyer, David, Genevieve Perron, and Mary Trudeau. (2005) *Analysis of Economic Instruments for Water Conservation*, submitted to Canadian Council of Ministers of the Environment, Water Conservation and Economics Task Group, Canadian Council of Ministers of the Environment, Winnipeg, Manitoba, pp1–82

40. Griffin, Kasandra, Greg Leventis, and Brian McDonald. (2010) *Implementing a Public Goods Charge*, UC – Berkeley, Goldman School of Public Policy Analysis Project, July 12, 2010

41. Florida Farm Bureau. (2008) "Issues & public policy: Policy book: Water/natural resources," www.floridafarmbureau.org/issues/policy_book/Water, accessed 11 July 2010

42. Department of Revenue, Washington State. (2010) "Bottled water," http://dor.wa.gov/content/findtaxesandrates/retailsalestax/bottledwater/, accessed 11 July 2010. The fee was repealed in late 2010, however.

43. Florida Department of Environmental Protection. (2009) *Water Bill Talking Points*

44. Florida Department of Environmental Protection. (2009) *$100 Million Scenarios for Water Use Fees*

CHAPTER THIRTEEN

Farming the Water

M ost Florida farms are relatively small, produce valuable commodities, and do not cause significant water resource problems. However, the conversion of wetlands to usable farm land was the main motivation for draining the Everglades and the huge marshes of the Upper St. Johns River, cutting off 20,000 acres of the wetland borders of Lake Apopka and channelizing many rivers for drainage and flood control. The boosters and land speculators who created Miami Beach, Golden Gate Estates, and other massive housing developments in wetlands had their own fleet of dredges, but they could not compete in scale with massive agricultural drainage operations funded by government agencies.

The effects of that agricultural heritage persist today. Industrial-scale agriculture in Florida often damages water quality and seriously impairs natural ecosystems while using amounts of water much higher than its economic contributions.

Addressing agricultural water problems is necessary for Florida to reach water resource sustainability. What are those current unsustainable patterns in industrial agriculture?

FARMING IS NUMBER ONE IN WATER USE

In Florida, agriculture withdraws more fresh water than any other activity. As noted in Chapter 10, Water Supply Is Easy, Florida withdraws more water to irrigate about 1.55 million acres of farm land than any other state east of the Mississippi River. The U.S. Geological Survey reports that during the relatively dry year of 2000, agriculture withdrew 48 percent of all of the freshwater in the state. Even during the relatively wet year of 2005, agriculture accounted for 40 percent of all freshwater (2.77 out of the 6.87 billion gallons a day totals).[1] Who irrigates? A very small number of farms. A total of only 349 farms in Florida have

64 percent of all of the acres irrigated in the state.[2] About 15 percent of agricultural irrigation is not for food at all but for ornamental plants and turfgrass.

Most of the water withdrawn for agricultural purposes is not returned to a surface water body or to groundwater where it would be available for another user. In agricultural irrigation, the majority of water withdrawn is evaporated into the air, stored in plants, or transpired by plants into the atmosphere. The result is that most irrigation water is not returned to the source and is instead "consumed." The amount *consumed* is even more significant than the amount only *withdrawn*.

The percent consumed in agricultural irrigation varies somewhat with rainfall because farms irrigate less, and consume less water, when rain is plentiful. In 2000, agriculture consumed 64.4 percent of all of the freshwater consumed in the state. In 2005, due to lesser need for agricultural irrigation that year, farming accounted for 62 percent of the water consumed in Florida.[3] The agricultural sector consumes more than three-fifths of all water consumed in Florida. Long-term water management in Florida has to focus on farm water use as a first priority.

THE BUSINESS OF FLORIDA FARMING

A very small number of large farms dominate Florida agriculture. The 2007 U.S. Census of Agriculture reports that the largest 2,067 farms account for 85 percent of the market value of agricultural production in Florida.[4] Table 13.1 shows census estimates for the number of farms accounting for 50 percent of market value for major Florida commodities.

Table 13.1 *Number of Farms Producing 50% of Sales by Commodity*

Commodity	Number of Farms Producing 50% of Farm Sales in 2007
Vegetables, melons, potatoes, and sweet potatoes	67
Fruits, tree nuts, and berries	113
Nursery, greenhouse, floriculture, and sod	107
Milk and other dairy products	32
Total farm sales	302 farms (only 0.7% of all farms)

The largest 349 irrigated farms accounted for 64 percent of all of the acres irrigated in the state for agricultural purposes (997,682 of a total of 1,552,118 acres). These 349 farms of 2,000 or more acres are large businesses, not Jeffersonian agrarians.

Farms use a great deal of land and water but are a surprisingly small part of the state's economy. Enterprise Florida, the state's primary organization for economic development, reports that the combined Agriculture, Forestry, and Mining sector of the state's economy is only 0.9 percent of the state total domestic product.[5] A more detailed estimate comes from the Bureau of Economic Analysis in the U.S. Department of Commerce. Within Florida, BEA estimates that the Agriculture, Forestry, Fishing, and Hunting sector of the economy had a total domestic product value of $5.5 billion in 2009, only 0.74 percent of the state's total of $737 billion.[6]

However, activities like Fishing and Hunting are not what is commonly thought of as "agriculture." Within what BEA calls the Crop and Animal Production (Farms) sector of the state's economy, it estimated a domestic product in Florida of $4.62 billion, which is only 0.6 percent of the state's total domestic product.

Much larger estimates of the economic impact of agriculture are published regularly. For example, the Florida Department of Agriculture often refers to agriculture as "the state's second largest industry."[7] This statement may be based on a misunderstanding of reports from the Institute of Food and Agricultural Sciences located at the University of Florida. The Institute publishes a biennial report on the economic impact of agriculture and "related industries."[8] They totaled up an economic impact in 2008 of nearly $163 billion, almost 1.4 million jobs, and 8 percent of the state's domestic product. However, the IFAS report title itself says it addresses much more than farming: "*natural resources, food and kindred product manufacturing and distribution, and service industries.*" For example, the total economic impact calculated in the IFAS report includes Florida oil and gas wells, soft drink manufacturing, golf courses, recreational fishing, bars, restaurants, and supermarkets. Thus, most of the economic impact in this total does not come from agriculture at all. The IFAS report does provide a total for farm activities, in what it calls the Crop, Livestock, Forestry, and Fisheries Production sector. That amounts to about 0.9 percent of the state's domestic product. Agriculture is a minor part of the state's economy, particularly in light of its enormous amount of water use.

WATER RESOURCE DAMAGE

Two-thirds of the 24 million acres of wetlands lost in the United States between 1954 and 2002 were due to agriculture.[9] Florida trends are similar: government drainage projects – organized primarily to facilitate agricultural activities – helped destroy Florida wetlands in the nineteenth and twentieth centuries, as discussed in Chapters 1 and 2. Even today, state law provides the industry a partial exemption from wetland regulation: "Nothing herein, or in any rule, regulation, or order adopted pursuant hereto, shall be construed to affect the right of any person engaged in the occupation of agriculture, silviculture, floriculture, or horticulture to alter the topography of any tract of land for purposes consistent with the practice of such occupation." (Section 373.406(2),F.S.) The 2011 Legislature widened this exemption significantly.

Nationwide, agriculture is the leading cause of water quality impairment in rivers and lakes as well as a major contributor to water quality problems in estuaries.[10] For some watersheds in Florida, agriculture is the primary polluter. For example, 98 percent of the phosphorus brought into the watersheds tributary to Lake Okeechobee is imported for agricultural operations.[11] The U.S. Environmental Protection Agency determined that current phosphorus discharges are so high that the lake will not meet federal water quality standards unless discharges of phosphorus are reduced by 76 percent. The upper St. Johns River has a similar pattern: agriculture accounts for 88 percent of human contributions of phosphorus to the system.[12]

Many Floridians put excessive amounts of fertilizer on their suburban lawns, but this activity is dwarfed by the amount of fertilizer used by farmers. Of the more than 2 million tons of fertilizer used in Florida on 3 million acres of land, three-quarters of the total are applied by farms.[13] Excessive fertilizer use is a source of significant water quality problems for the state's surface waters and groundwaters.

FARMING THE POLITICAL SYSTEM

Perhaps the most valuable Florida agricultural crop is of politicians disposed favorably toward farming. The Commissioner of Agriculture is one of only four statewide elected officials in Florida, although rural counties have less than 7 percent of the state's population.[14] In 2007 Commissioner Charles Bronson said with pride, "My department, the Florida Department of Agriculture and Consumer Services, is the largest state department of agriculture in the country with over 3,700 employees."[15] Agricultural interests demonstrated their power in the Florida legislature during the 1972 "Year of the Environment," which included the creation of the modern water management system. Earl Starnes, first director of the state Division of State Planning and a long-time board member of the Suwannee River Water Management District, looked back with some regret on how agriculture was exempted from the growth management reforms of the time:

> Q.: So you guys create these four [sets of growth management] recommendations, the Governor takes them to the Legislature, and the Legislature passes the whole package in 1972.
>
> Starnes: The whole package, with one reservation: that agriculture was excluded from the definition of development. There is a very long complicated definition in the statute, Chapter 380, and it specifically says what development means in Florida under Florida law. It [originally] included agriculture [but] in order to get it out of the Senate, it was necessary to exclude agriculture.[16]

That level of political influence has not diminished. For several years, the senior vice president of operations for the U.S. Sugar Company served as a member of the governing board of the South Florida Water Management District. He resigned to avoid the "appearance of a conflict" of interest when the state proposed in 2008 to buy out the company.[17] Nor was sugar the only major farm industry represented on governing boards. In 2009, for example, the Florida senate confirmed other agricultural appointments to governing boards, including the past chair of the Florida Fruit and Vegetable Association, the executive vice president of the Indian River Citrus League, an employee of the largest firm managing citrus groves in the state, the executive director of a regional nursery wholesale group, and the treasurer of the Florida Cattleman's Association.[18] In early 2011, four of the chairs of governing boards of the five water management districts owned agricultural businesses.

"Big Sugar's" political power reached into the White House. On the morning of February 19, 1996, Vice President Al Gore announced new Clinton administration initiatives to restore the Everglades, including the president's support of the proposed penny-a-pound tax on sugar to help in restoration funding.[19] That same afternoon, President Bill Clinton met with Monica Lewinsky to explain that he wanted to put a stop to their relationship, which threatened his marriage and his presidency. Despite the importance of this conversation, Clinton interrupted their talk in order to take a 22-minute call from the co-owner of the Flo-Sun sugar company in the Everglades Agricultural Area.[20] Florida voters, after an intense and very expensive political campaign, accepted the arguments of the sugar companies and did not put the penny-a-pound tax in the state constitution.

FARMING THE TAX CODE

Florida farmers benefit from a range of national agricultural policies. For example, farmers can buy crop insurance sponsored by the U.S. Department of Agriculture. From 1995–2009, the insurance system cost the federal government $1.17 billion more than the $373 million collected in Florida.[21] Powerful incentives for agricultural irrigation are also deeply embedded in the state's own property tax system. In 1949, at a critical turning point in Florida water management history, agricultural interests devised an ingenious way to shift to others most of the costs of running the new Central and Southern Florida Flood Control District. Jeanne Bellamy of the *Miami Herald* (and later South Florida Water Management District governing board member) described an important meeting in the city of Okeechobee where the district's tax structure was designed to benefit agriculture:

> 'Millard Caldwell and Fuller Warren appointed a committee,' she recalled. 'I was secretary. It was suggested that the tax rate be on a sliding scale, with the lowest in Miami Beach, because of minimum benefits there; the maximum benefits would be in the agricultural area. But the word was passed at Okeechobee – there would be a flat millage or no district.' They got their wish. A uniform millage was authorized by the Legislature when it created the District.[22]

Because agriculture paid less than the benefits it received from the Central and Southern Florida Flood Control Project, Florida also got more agricultural pollution than it wanted. The state property tax system contains an even more important tax break for agriculture. The Florida legislature directs that farms are to be assessed for property taxes at much less than actual value ("use" value rather than market value). If a farm could be sold for a million dollars for a housing development but is worth only $150,000 if engaged in farming, the law says that it is to be treated for tax purposes as if it were worth only $150,000. As a result, agriculture statewide pays only 14.2 percent of the taxes it would if it were assessed at the actual value of the property. That special provision cost local and regional governments $1.27 billion in tax revenues in 2008.[23]

Proponents of the tax break defend it by saying that it keeps agricultural land in production. However, many other states also interested in supporting agriculture do not provide such unrestricted tax breaks to agriculture. The *2007 Florida Tax Handbook*, prepared by the legislature's Office of Economic and Demographic Research, listed seven different possible ways to lessen or recover the "tax expenditure" for an agricultural property tax break from "farmers" who turn out to be developers more than farmers. None of these reforms is in Florida law. As tax policies stand now, farmers are given financial incentives to use more water because their water-using activities are exempt from some major taxes.

FOODS FROM FLORIDA

The largest food problem of America is obesity. The December 2010 national "Dietary Guidelines" emphasize that the condition of being overweight or obese is "dramatically higher than it was a few decades ago" and that this increases the "risk of many health problems."[24] The Guidelines also explain:

> Strong evidence shows that eating patterns that are low in calorie density improve weight loss and weight maintenance, and also may be associated with a lower risk of type 2 diabetes in adults.

Foods that are low in calorie density have low calories per unit of food weight. The two largest food producing activities in Florida are sugarcane and citrus fruits (primarily for juice). Both of these are relatively "calorie dense" foods and the health implications are discussed in sections below.

A SWEET DEAL FOR SUGAR

Without irrigation, the sugarcane industry in the Everglades Agricultural Area could not exist. Water use there is phenomenally large. The 875 million gallons a day used in sugarcane farming in 2005 was one-eighth of all freshwater used in Florida.[25] Sugar farming uses enormous amounts of water for relatively small economic value: about 5,000 gallons for each dollar of economic output.[26] Many of the water resource problems of South Florida would be alleviated if government policies did not subsidize the sugar industry and its use of water. Federal policies have enabled the industry to prosper, according to the Congressional Budget Office:

> The federal government protects the price of sugar by restricting imports, making below-market-rate loans available to processors, and limiting the amount of sugar that processors can sell domestically. Because of that support, domestic growers' supply for the national sugar market increased from roughly 55 percent of the total prior to the early 1980s to 86 percent in 2005.[27]

Enormous subsidies and import protections encourage sugar farming and its associated use of water. The International Trade Administration of the U.S.

Department of Commerce found in 2006 that the sugar program cost the national economy about a billion dollars a year.[28] They found that for each sugar growing and harvesting job fostered by protective trade policies in the United States, close to three other jobs are lost elsewhere in the national economy due to artificially high sugar prices.

Sugar politics cause these economic losses but they also impose health risks. The new national Dietary Guidelines point out that added sugar contributes "an average of 16 percent of the total calories in the American diet" and "should be reduced."[29]

BIG CITRUS

The new Dietary Guidelines report that Americans from 2 to 30 years old consume more than half of the fruit they consume as juice. This matters because,

> Although 100% fruit juice can be part of a healthful diet, it lacks dietary fiber and when consumed in excess can contribute to extra calories. The majority of the fruit recommended should come from whole fruits, including fresh, canned, frozen, and dried forms, rather than from juice.[30]

The Florida Citrus Commission reports that over 85 percent of citrus in the state is processed and sold as juice.[31]

Making that much juice involves huge amounts of irrigation of fruit trees. The citrus industry uses even more water than sugarcane in a thirsty state. Harriet Beecher Stowe, the author of *Uncle Tom's Cabin*, called the orange tree the "noblest" tree which "the Lord God caused to grow eastward in Eden."[32] From a botanical perspective, however, citrus plants are non-native introduced species, grown as a monoculture and managed for profitability like any other crop plant. Citrus has the highest herbicide use of any crop in Florida and the third highest rate of insecticide use.[33] Nutrient pollution is also a problem, especially for the large citrus groves grown on low-nutrient sandy soils. Citrus operators fertilize their groves at least three to four times a year with plant nutrients that usually move freely through porous soils.[34] Citrus is grown on 576,000 acres of Florida. Water use is correspondingly large. In the dry year of 2000, citrus water use was 1.82 billion gallons a day, amounting to 22 percent of all of the water use in Florida. Wetter conditions in 2005 resulted in less water use (996 million gallons a day), which was still 14.5 percent of the state's total.

Orange growers persuaded Congress to erect a variety of trade barriers against imports. The appraisal of the U.S. Sugar properties in the South Florida Water Management District also covered this topic because the "sugar" company also operates 22,000 acres of orange groves. The appraisal reported that the import tariff was 7.85 cents per liter of orange juice. Like sugar tariffs, citrus tariffs improve grower profits but are paid ultimately by consumers. Those same tariffs that impose higher costs on consumers also encourage higher water use.

In regard to greenhouse gas emissions, oranges are especially energy-intensive: crops like corn and soybeans return several calories of food energy for each calorie of fossil energy invested in their growth, but orange groves produce only about one unit of food energy for each unit of fossil energy used in their production.[35] The PepsiCo Corporation (which owns the Tropicana orange juice brand) estimates that the carbon footprint of a half-gallon of orange juice was 3.8 pounds of carbon dioxide.[36] High energy demands and high water use are key features of the citrus industry.

SACRED COWS?

The phosphorus in each cow's daily manure and urine is the same as that of 18 to 20 people.[37] The nearly 2 million large livestock animals in Florida generate the waste equivalent of 31 to 35 million residents. Concentrated livestock operations, such as dairies, have concentrated water quality problems. Florida dairies had 120,000 cows in 2008. Hot Florida summers reduce milk production per dairy cow by about a quarter from that in more moderate climates. The various health problems of dairy cows in Florida's hot climate result in about 35 percent of the animals being culled each year.[38] An individual high producing dairy cow can ingest 100 pounds of feed and 50 gallons of water every day, while also producing 195 pounds of manure and urine. The solid waste from Florida's dairy cattle alone is the equivalent of 2.1 to 2.3 million people. Dairy cows also are significant sources of methane, which is a greenhouse gas many times more powerful than carbon dioxide. Milk production is remarkably energy-intensive, using about 14 calories of fossil energy for every calorie of milk protein produced.[39] Each gallon of milk production is associated with the emission of about 15−21 pounds of greenhouse gases.[40]

The dairy business is run by a very small number of farmers. There were only 143 dairy farms in Florida in 2009.[41] Milk production in Florida depends upon a variety of federal programs of baroque complexity. For example, federal dairy policy discourages milk shipments between regions, even if consumers would pay less for milk if such transfers were allowed.[42] In addition to milk production subsidies, Florida tax dollars pay directly for the cleanup of dairy pollution. Florida dairy farmers have often been paid to stop polluting rivers, lakes, and aquifers rather than paying for pollution abatement themselves. For example, in the 1980s dairies in the Kissimmee watershed north of Lake Okeechobee were found to be major polluters of Lake Okeechobee. The Florida legislature appropriated funds to pay dairy operators $602 for every cow removed from the watershed and not replaced with another cow. The buyout paid for the removal of 14,039 cows.[43]

The state placed no restrictions on where in Florida dairy production could move after being paid to move out of the Lake Okeechobee basin. Unconstrained, some of the dairy operators relocated their cows into the vulnerable Suwannee River watershed. More than a quarter of all dairy cows in Florida are now located in four counties that border the Suwannee River (Gilchrist, Lafayette, Suwannee, and Levy). The karst geology of the Suwannee River region is one of the least desirable locations possible for intensive agricultural operations because pollutants are rapidly transported in that region's porous geology from the land surface into springs, rivers, and drinking water wells. As a result, dairy farms in the Suwannee River watershed are major contributors to the problem of nitrogen pollution.[44] The state is now paying again to help clean up these operations, through financial assistance to the Suwannee River Partnership.

WILL BIOFUELS BE DONE RIGHT?

The growing biofuel industry, if managed unwisely, will recreate all of the old Florida problems of taxpayer-subsidized destruction of water resources. The Energy Independence and Security Act of 2007 requires the production of 36 billion gallons of renewable fuels by 2022, including 16 billion gallons from "advanced" biofuels. If this is done right, Florida may become a major supplier of renewable energy and establish a stable revenue stream for agriculture. If done thoughtlessly, Florida will establish a new agricultural sector that wastes gargantuan quantities of water, demands enormous taxpayer subsidies, and damages the state's vulnerable ecosystems.

Would Biofuels Actually Reduce the Emission of Greenhouse Gases?

The potential benefits from growing biofuel crops must be compared to the energy inputs used in growing the feedstock because farming is energy intensive. About 107 gallons of oil equivalent is used to grow one acre of corn.[45] Lifecycle studies of energy crops usually find at least modest greenhouse gas benefits. Sugarcane, for example, apparently has net greenhouse benefits when used as a feedstock for a refinery making ethanol. The same may be true for cellulosic feedstocks (like sugarcane stalks and wood chips) if it proves possible to produce ethanol from them at commercial scales.

However, troubling recent studies point to different results if the lifecycle effects of expanding crop acreage are included in the calculations.[46] To grow the new energy crops in addition to current food crops, farmers necessarily must convert current forests or pasture lands to more intensive biofuel production. That conversion process causes an immediate large release of carbon to the atmosphere. Even if the new energy crop has annual greenhouse gas emission benefits, there will still be an immediate carbon debt to overcome. New corn acreage takes 167 years to overcome that initial debt. Even switchgrass, a possible alternative biofuel, can take 52 years to overcome the initial release of carbon.

Are Government Subsidies Well Designed?

Biofuel production has secured enormous government subsidies. The single largest energy tax expenditure in the U.S. budget is the tax credit for ethanol fuels, with a 5-year federal revenue cost from 2005–2009 of $17 billion.[47] In late 2010, as part of an enormous federal tax deal, the subsidy rate for ethanol of 45 cents for every gallon of ethanol was extended another year. The subsidy is given to refiners regardless of the amount of fossil fuel used in ethanol production, whether the feedstock is corn or cellulosic, the degree of water use efficiency, or the net effect on greenhouse gas emissions. Four-fifths of economists believe that the current ethanol subsidy should be reduced.[48]

Corn, Sugarcane, Alcohol?

The scale of the ethanol dream is breathtaking. Vast acreages of the United States and Florida may become feedstocks for ethanol and other liquid fuels, mostly harvested on giant industrial-scale farms. The state's climate may lead it to focus on sugarcane for energy conversion. Even sugarcane would require immense amounts of land and water. The land requirements for producing enough sugar-based ethanol to substitute for even 20 percent of the gasoline used in Florida would amount to 3,860 square miles – about 7 percent of the land mass of the state.[49] Floridians do not have that land to spare.

If conventional crops like corn and sugarcane cannot meet energy needs, have questionable greenhouse gas benefits, or have excessive environmental costs, perhaps the answer lies in growing woody crops and using new technologies to convert their cellulose to ethanol. In Florida, native species like switchgrass (*Panicum virgatum*) are being considered as biofuels. However, even these may result in massive amounts of new irrigation. Although switchgrass and some other potential energy crops can be grown on farms without irrigation, yields are larger with supplemental water.

Any serious effort to grow biofuels in Florida must also prevent undesired biological invasions from new species. That task is challenging because the ideal cellulosic biofuel doesn't need much fertilizer or pesticides, outcompetes weeds in seasonal growth, is not bothered by drought, and grows very rapidly. That description of an ideal biofuel crop is also a very good description of an invasive plant pest.[50]

How Much Water Might be Used?

Ethanol refineries require large amounts of water. About 3 to 4 gallons of water are used in the production facility for each gallon of ethanol.[51] The capital-intensive manufacturing plants, however, are only a small part of the ethanol production system. Water use in growing the feedstocks can be much larger. A national study estimated that ethanol produced by irrigated corn can require up to 324 gallons of water for each gallon of ethanol. Another way of expressing that is gallons of water per mile of automobile travel. One estimate for corn ethanol is that about 28 to 36 gallons of water is used for every mile traveled in a car fueled by that supply.[52] About a thousand gallons of water are needed to produce a single gallon of ethanol from sugar (based on the use of 72 gallons of water for each pound of sugar and using 14 pounds of sugar for each gallon of ethanol.[53]).

On a statewide scale, biofuels like switchgrass may become one of the state's biggest users of water. Without any irrigation, switchgrass may consume from 2 to 10 gallons of water for each gallon of eventual ethanol. If irrigation is used to increase production of cellulosic ethanol from feedstocks such as switchgrass, the amount of water used would be much higher.[54] Displacing only 20 percent of current Florida gasoline use with biofuels would require increasing total statewide irrigation by 360 percent if it were done with corn, and by 60 percent if done with sorghum.[55] Whatever crop is used, massive biofuel production could mean massive increases in agricultural water use.

Biofuels are Not Magic

A biofuel path makes sense only if it is certified to reduce lifecycle greenhouse gas emissions, has tax subsidies that are proportional to the net greenhouse effects of different crops, protects water and other natural resources from damage, and sustains local farmers and communities. Biofuel production will not achieve these policy goals unless Floridians make conscious choices for sustainability.

TOWARD SUSTAINABLE AGRICULTURE IN FLORIDA

Florida agriculture is a business, not a virtue,[56] and that business enjoys broad public support. Florida needs a strong and productive agricultural sector that supports vibrant agricultural communities while sustaining natural systems. A confluence of trends is emerging that may transform Florida agriculture. The old assumptions are being challenged:

- *Soils can be mined.* Florida lost immense quantities of organic soils in the nineteenth and twentieth centuries, such as several feet of irreplaceable organic soils in the Everglades Agricultural Area. Intensive agricultural practices that mine soil to produce food and fuels are not sustainable because they cannot last indefinitely.
- *Fossil fuel is cheap.* About 17 percent of fossil fuel in the United States is devoted to food production.[57] An agricultural industry built on cheap oil will lose a competitive edge to other types of agriculture that are not so dependent on fossil fuels
- *Water resources can absorb unlimited wastes.* When Florida was an empty state in pioneer days, natural resources were abundant and humans were few. It may have made economic sense back then to treat water resources as bottomless waste receptacles.[58] Today, the state has close to 19 million residents and its water resources have little ability to absorb more pollution. Agricultural water pollution cannot be disregarded.
- *Water supplies are abundant.* Large-scale irrigation operations dominate in Florida agriculture. This practice was understandable when there was little competition from other uses, sufficient water for all, limited understanding of ecosystem needs, and no prospect of increased climatic variability. It makes less sense in the twenty-first century with growing competition for limited supplies of inexpensive water.
- *Irrigated agriculture has first claim on the available supplies of water.* As discussed earlier in this chapter, irrigated agriculture provides only small economic

returns to the state in proportion to the amount of water it uses. When water use is a business decision, Florida has to be sure that the limited available supplies of water generate the highest economic return.

- *Climate is stable.* As discussed in the next chapter, Florida's climate is changing. Different patterns of precipitation are expected, including more common or more severe droughts. These changes will pose serious challenges to irrigated agriculture.
- *Greenhouse gas emissions can be ignored.* Nationally, 10 calories of fossil energy are expended in the production process for every calorie of American food. Even at the farm level, three times as much fossil energy is used as is eventually taken off the farm in food energy.[59] Florida agricultural practices, in particular, release enormous amounts of greenhouse gases by farming high-organic soils. To meet the challenge of reducing global greenhouse gas emissions, Florida farming will have to make major adjustments.

The old assumptions, if they ever were true, are true no longer. A sustainable agriculture system in Florida would mean more concern for the long-term health of water resources, the land, and the people they support. This new model of sustainable Florida agriculture would mean that farms would use much less water and become less energy intensive than they are today. They will have to operate more as farm ecosystems than as enormous factories for food and fuel. They would benefit from fostering diverse plant and animal production rather than raising monocultures of corn, oranges, sugarcane, or dairy cows. United States agricultural policy would make payments primarily to farmers who protect ecosystem services rather than using subsidies that distort markets and impair economic efficiency. John Ikerd, a leader in the sustainable agriculture movement, advises that we must ask of every agricultural activity, is it "ecologically sound, economically viable, and socially responsible?"[60] Wise answers to those questions would mean that Florida agriculture would protect vulnerable water resources from enduring harm.

NOTES

1. Marella, 2009.

2. U.S. Department of Agriculture, National Agricultural Statistics Service. (2008) *2007 Census of Agriculture*, Florida Data, Table 10, "Irrigation: 2007 and 2002"

3. Personal communication from Rich Marella, U.S. Geological Survey

4. U.S. Department of Agriculture. (2008) *2007 Census of Agriculture*, vol. 1, chap. 2: State Level Data, Table 2. "Market Value of Agricultural Products Sold Including Direct Sales: 2007 and 2002"

5. Enterprise Florida, Inc., Marketing and Strategic Intelligence Division. (2008) *The Florida Economy*, Enterprise Florida, Inc., Orlando, FL

6. U.S. Bureau of Economic Analysis. (2009) "Gross domestic product by state," http://www.bea.gov/regional/gsp/, accessed February 8, 2010. This estimate of the contribution made to the state's overall domestic product from "Agriculture, forestry, fishing, and hunting" was even lower than the 2007 estimate of 0.9 percent.

7. Florida Department of Agriculture. (2003) *Florida's Agricultural Water Policy*, Florida Department of Agriculture, Tallahassee, FL, p 2

8. Hodges, Alan W., and Mohammad Rahmani. (2010) *Economic Contributions of Florida Agriculture, Natural Resources, Food and Kindred Product Manufacturing and Distribution, and Service Industries in 2008*, EDIS Document FE829, Food and Resource Economics Department, Institute of Food and Agricultural Sciences, University of Florida, Gainesville, FL

9. Hansen, LeRoy. (2008) *Wetlands Status and Trends*, Agricultural Resources and Environmental Indicators, Economic Research Service/USDA 2006 ed/EIB-16

10. Ribaudo, Marc, and Robert Johansson. (2006) *Water Quality: Impacts of Agriculture*, Agricultural Resources and Environmental Indicators, Economic Research Service/USDA 2006 ed/EIB-16, p33

11. U.S. Environmental Protection Agency, Region 4. (2008) *Final Total Maximum Daily Load (TMDL) for Biochemical Oxygen Demand, Dissolved Oxygen, and Nutrients in the Lake Okeechobee Tributaries Osceola, Polk, Okeechobee, Highlands, Glade, and Martin, Florida*, p41

12. Gao, Xueqing. (2006) *TMDL Report: Nutrient and DP TMDLs for the St. Johns River above Lake Poinsett (WBID 2893L), Lake Hell n' Blazes (WBID 2893Q), and St. Johns River above Sawgrass Lake (WBID 2893X)*, Florida Department of Environmental Protection, Tallahassee, FL, p30

13. Florida Department of Agriculture, Bureau of Compliance Monitoring. (2010) "Historical total fertilizer summary," www.flaes.org/complimonitoring/index.html, accessed 3 July 2010

14. U.S. Department of Agriculture, Economic Research Service. (2010) *Florida Fact Sheet*, July 1, 2010

15. Bronson, Charles H. (2007) Testimony before the Subcommittee on Horticulture and Organic Agriculture, Committee on Agriculture, United States House of Representatives, October 3, 2007

16. Oral History Program. (2000) Earl Starnes interview, "Florida growth management," 17 June

17. Wade, Malcolm (Buddy) Jr. (2008) Letter to Governor Charlie Crist, June 24, 2008

18. Florida Senate Committee on Environmental Preservation and Conservation. (2009) Meeting Packet, March 17, 2009

19. Godfrey, Matthew C., and Theodore Catton. (2006) *River of Interests: Water Management in South Florida and the Everglades, 1948–2000*, Chapter Six, U.S. Army Corps of Engineers, Washington, DC, p329

20. Office of the Independent Counsel. (1998) Referral to the United States House of Representatives pursuant to Title 28, United States Code, § 595(c), September 9, 1998 (the "Starr Report")

21. EWG Farm Subsidy Database, Crop Insurance in Florida, http://farm.ewg.org/cropinsurance.php?fips=12000&summpage = SUMMARY Accessed January 31, 2011

22. Huser, Thomas. (1990) *Into the Fifth Decade: The First Forty Years of the South Florida Water Management District, 1949–1989*," South Florida Water Management District, West Palm Beach, FL p11

23. Office of Economic and Demographic Research, Florida Legislature. (2008) *2008 Florida Tax Handbook*, p151. This total fell in 2010 to about $1 billion due to the general decline in Florida property values

24. U.S. Department of Agriculture and U.S. Department of Health and Human Services. Dietary Guidelines for Americans, 2010. 7th Edition, Washington, DC: U.S. Government Printing Office, December 2010

25. Personal communication from Rich Marella, U.S. Geological Survey

26. Blackhurst, Michael, Chris Hendrickson, and Jordi Sels i Vedal. (2010) "Direct and indirect water withdrawals for U.S. industrial sectors," *Environmental Science and Technology,* vol. 44, no. 6, pp2126–2130

27. Congressional Budget Office. (2006) *How Federal Policies Affect the Allocation of Water,* p18

28. U.S. Department of Commerce, International Trade Administration. (2006) *Employment Changes in U.S. Food Manufacturing: The Impact of Sugar Prices*

29. Dietary Guidelines, p.9

30. Ibid., p. 36

31. Florida Citrus Commission "Florida Citrus Facts" http://www.floridajuice.com/juice. php Accessed on February 2, 2011

32. Stowe, Harriet Beecher. (1873, 1968) *Palmetto Leaves,* 1968 facsimile reproduction of the 1873 ed, University Press of Florida, Gainesville, FL, p18

33. Choquette, A. F., and Kroening, S. E. (2009) *Water Quality and Evaluation of Pesticides in Lakes in the Ridge Citrus Region of Central Florida,* U.S. Geological Survey Scientific Investigations Report 2008-5178

34. National Science Foundation Center for Integrated Crop Management. (2010) "Florida citrus timeline, crop production," http://pestdata.ncsu.edu/croptimelines/pdf/FLCitrus.pdf, accessed 25 March 2010

35. Pimentel, David. (2006) "Impacts of organic farming on the efficiency of energy use in agriculture," *Organic Center State of Science Review,* Cornell University, Ithaca, NY, August 2006

36. Pepsico Corporation, "A big step toward a reduced carbon footprint", http://www.pepsico.com/annual09/ourStories_A_Big_Step_Toward_Reduced_Carbon_ Footprint.html, accessed January 28, 2011

37. Novotny, Vladimir. (2002) *Water Quality: Diffuse Pollution and Watershed Management,* John Wiley & Sons, Inc., New York, p536

38. Giesy, Russ, Albert de Vries, and Jay Giesy. (2008) *Florida Dairy Farm Situation and Outlook 2008,* Document AN200, Institute of Food and Agricultural Sciences, University of Florida, Gainesville, FL

39. Pimentel, David, Sean Williamson, Courtney Alexander, Omar Gonzalez-Pagan, Caitlin Kontak, and Steven Mulkey. (2008) "Reducing energy inputs in the U.S. food system," *Human Ecology,* 36, pp459–471

40. Thoma, Greg et al., "Regional Analysis of Greenhouse Gas Emissions from Milk Production Practices in the United States" University of Arkansas and Michigan Technological University. Innovation Center for U.S. Dairy, October 1, 2010, pp11

41. Florida Department of Agriculture. (2010) "Florida dairy facts," www.doacs.state.fl.us/ dairy/dairyfacts.htm, accessed 29 March 2010

42. Balagtas, Joseph V.(2007) "U.S. dairy policy: Analysis and options," in *Agricultural Policy for the 2007 Farm Bill and Beyond,* American Enterprise Institute, Washington, DC

43. National Center for Environmental Economics. (1993) "Economic incentives for pollution control," in *Geographic Incentives,* Section 7.4.12.1, State Initiatives, National Center for Environmental Economics, U.S. Environmental Protection Agency, Washington, DC

44. Cabrera, Victor E. (2004) "Modeling north Florida dairy farm strategies to alleviate ecological impacts under varying climatic conditions: An interdisciplinary approach," Ph.D. dissertation, University of Florida, Gainesville, FL

45. Pimentel, David, quoted in Myers, Norman and Jennifer Kent. (2005) *The New Atlas of Planet Management,* University of California Press, Berkeley, CA, p22

46. Searchinger, Timothy, Ralph Heimlich, R. A. Houghton, Fengxia Dong, Amani Elobeid, Jacinto Fabiosa, Simia Tokgoz, Dermot Hayes, and Tun-Hsiang Yu. (2008) "Use of

U.S. croplands for biofuels increases greenhouse gases through emissions from land use change," *Science*, vol. 319, no. 5867, pp1238–1240

47. Cox, Craig, and Andrew Hug. (2010) *Driving Under the Influence: Corn Ethanol and Energy Security*, Environmental Working Group, Washington, DC

48. Whaples, Robert. (2009) "The policy views of American Economic Association members: The results of a new survey," *Econ Journal Watch*, vol. 6, no. 3, pp337–348

49. Cohen, Matthew, Jason E. Evans, and Mark T. Brown. (2008) *The Water Resource Implications of Large-Scale Bioethanol Production*, Water Institute 2007 Program Initiation Report, University of Florida, Gainesville, FL, pp3–4

50. Marland, Gregg, and Michael Obersteiner. (2008) "Large-scale biomass for energy, with considerations and cautions: An editorial comment," *Climatic Change*, vol. 87, no. 3-4, pp335–342

51. Aden, Andy. (2007) "Water usage for current and future ethanol production," *Southwest Hydrology*, September/October, pp22–23. Another study found that the "dry mill" process used about 3.45 gallons of water per gallon of ethanol and the "wet mill" process used about 3.92 gallons of water per gallon of ethanol produced: Wu, May. (2008) *Analysis of the Efficiency of the U.S. Ethanol Industry 2007*, Center for Transport Research, Argonne National Laboratory, Argonne, IL

52. King, Carey W., and Michael E. Webber. (2008) "Water intensity of transportation," *Environmental Science and Technology*, vol. 42, no. 21, pp7866–7872

53. Erickson, John. (2008) "Feedstock for bioenergy in Florida," presentation at the Florida Farm to Fuel Summit, University of Florida, Gainesville, FL, 2008

54. Wu, M., M. Mintz, M. Wang, and S. Arora. (2009) *Consumptive Water Use in the Production of Ethanol and Petroleum Gasoline*, ANL/ESD/09-1, Center for Transportation Research, Energy Systems Division, Argonne National Laboratory, Argonne, IL

55. Cohen et al., op. cit., note 43

56. The phrasing that agriculture businesses really are businesses and not undiluted virtues comes from Rick Bastasch's assessment of water problems on the Pacific coast in *Oregon Water Handbook: A Guide to Water and Water Management* (2006), Oregon State University Press, Corvallis, OR

57. Chivian, Eric, and Aaron Bernstein. (2008) *Sustaining Life: How Human Health Depends on Biodiversity*, Oxford University Press, New York, p357

58. Beddoe, R., Costanza, R., Farley, J., Garza, E., Kent, J., Kubiszewski, K., Martinez, L., McCowen, T., Murphy, K., Myers, N., Ogden, Z., Stapleton, K., and John Woodward. (2009) "Overcoming systemic roadblocks to sustainability: The evolutionary redesign of worldviews, institutions, and technologies," *Proceedings of the National Academy of Sciences*. vol. 106, no. 8, pp2483–2489

59. Ikerd, John E. (2008) *Crisis and Opportunity: Sustainability in American Agriculture*, University of Nebraska Press, Lincoln, NE

60. Ikerd, ibid

CHAPTER FOURTEEN

Turning Up the Thermostat in Paradise

*I*n the nineteenth-century, settlers in Florida often looked upon water resources as hellish obstacles to development. Watery areas were infested with mosquitoes, blocked by wide expanses of impassable marshes, and frustratingly difficult to drain. Views have changed substantially since then. Even with all of the damage done to its water resources, Florida still is perceived today as a kind of earthly paradise – in large part because of its water. Recent books on Florida natural resource management adopt this theme, such as *Paving Paradise* (Craig Pittman and Matthew Waite, 2009), *The Swamp: The Everglades, Florida, and the Politics of Paradise* (Michael Grunwald, 2006), *Paradise Lost? The Environmental History of Florida* (Jack Davis and Raymond Arsenault, editors, 2005), *Journeys through Paradise* (Gail Fishman, 2001), and *Some Kind of Paradise* (Mark Derr, 1998).

However, global climate change seems poised to convert Florida's Eden into a hotter place. The atmospheric effects of greenhouse gases threaten, in this or the next century, to melt enough of the polar ice caps to raise sea levels by several feet or more, disproportionately affecting places with low elevations like Florida. Climate change may drown Florida's coastal wetlands, including major sections of the Everglades. Rising temperatures will likely also mean many other changes such as different rainfall patterns, changes in hurricane strength or frequency, and the loss of coral reefs. These, too, would have profound consequences on Florida. Climate change is the only existential threat to what Chapter 1 calls the "watery foundation" of the state. Florida citizens and water managers will be facing difficult decisions on how best to respond to a changing climate. It will be necessary both to reduce greenhouse gas emissions and make adaptations to whatever level of climate change is set in motion.

This is not news. As far back as 1984, the *Water Resources Atlas of Florida* noted that sea level rise induced by global warming would cause "increased saltwater intrusion into fresh water aquifer systems" and that "the frequency and distribution of floods, droughts, and severe storms could be much different in the next century."[1] Today the connections between climate change, water, energy production, and greenhouse gases are better understood than in 1984. That higher level of understanding is a result of what John Holdren, President Obama's Science Adviser and the former president of the American Association for the Advancement of Science, calls the "largest, longest, costliest, most international, most interdisciplinary, and most thorough formal review of a scientific topic ever conducted."[2] The reality of human-caused climate change is now very well established, including its significance for water management.

FACING UP TO CLIMATE CHANGE IN FLORIDA WATER MANAGEMENT

You don't have to be a meteorologist to see which way the wind is blowing, and you don't have to be a climatologist to conclude that human activities are intensifying the greenhouse effect for the earth and for Florida. Our civilization is adding vast quantities of greenhouse gases to the atmosphere by burning fossil fuels, at the unsustainable pace of a million year's worth of fossilized plants a year. Human activities release much more carbon dioxide to the atmosphere today than do volcanoes.[3] Carbon dioxide and methane are the greenhouse gases of greatest concern, both of which are associated with human activities. Florida alone added 337 million metric tons of carbon dioxide in 2005. Global fossil fuel combustion causes the concentration of carbon dioxide in the atmosphere to increase about 2 parts per million (ppm) per year.[4] Before the industrial age began, the atmospheric concentration of carbon dioxide (CO_2) was about 280 ppm. By 2010, the annual average concentration had risen to 390 ppm, higher than at any point in the last 800,000 years.[5]

Not surprisingly, the planet's temperature has risen as greenhouse gases have increased so dramatically. Many lines of scientific investigation have confirmed and amplified on this basic physical phenomenon. The Intergovernmental Panel on Climate Change affirmed this conclusion in each of its 5-year assessments since 1989. In 2006 the American Association for the Advancement of Science stated that the "scientific evidence is clear: global climate change caused by human activities is occurring now, and it is a growing threat to society" and that "the longer we wait to tackle climate change, the harder and more expensive the task will be."[6] The national academies of science of 13 major industrialized countries also made a clear statement in 2005 on the need for action:

> It is unequivocal that the climate is changing, and it is very likely that this is predominantly caused by the increasing human interference with the atmosphere. These changes will transform the environmental conditions on Earth unless counter-measures are taken. Our present energy course is not sustainable.[7]

Other eminent scientific organizations continue to affirm the same basic conclusions. For example, on October 21, 2009, 18 American scientific organizations wrote to the U.S. Senate to state that "climate change is occurring" and that "rigorous scientific research demonstrates that the greenhouse gases emitted by human activities are the primary driver." They added that if society is to "avoid the most severe impacts of climate change, emissions of greenhouse gases must be dramatically reduced."[8]

Roger Revelle and Hans Suess observed more than a half-century ago that putting so much carbon in the planet's atmosphere is a massive "geophysical experiment."[9] In effect, humans are turning up the thermostat in the atmosphere of Paradise. Without very large reductions in the emission of greenhouse gases, the average global temperature may increase by 6 °C by 2100. Nor would the effects of increasing greenhouse gases end in 2100 or even stabilize then because greenhouse gases stay in the atmosphere for so long after being placed there. Even after 10,000 years, 10 percent of the CO_2 that humans place in the atmosphere will still be present and still exerting a greenhouse effect.[10] As a result, the factors causing the changes by 2100 will continue to raise sea level for centuries after that date. A study from the National Research Council warns that current emissions of greenhouse gases could "lock the Earth and many future generations of humans into very large impacts, some of which could become very severe."[11]

It is worrying that the pace of climate change appears to be exceeding scientific projections. The 2007 report from the Intergovernmental Panel on Climate Change (IPCC) is the gold standard of the science of climate change but it appears to have underestimated the speed of planetary change. The rate of sea level rise over the past 15 years is 3.4 cm/decade − 80 percent more than IPCC predictions.[12] A hotter planet for Florida will mean, at a minimum, altered rainfall patterns, a rising sea level, ocean acidification, movement of invasive species, and serious water supply challenges. Despite these threats, Florida has taken only a few steps to reduce its contributions to climate change and to adapt to climate changes already under way. Why is that? One factor might be a general mistrust of science nationally and in Florida. This attitude extends to evolutionary theory as well as climate change. A 2008 *St. Petersburg Times* poll found that only 22 percent of Floridians want public schools to teach an evolution-only curriculum, while 50 percent want only faith-based theories such as creationism or intelligent design.[13] A state electorate unwilling to embrace the scientific facts of evolution may also be likely to deny the facts of climate change, including the state's water resource vulnerability.

Another reason why Floridians and Americans have not done much about climate change is the disproportionate attention given to a small minority of scientific dissenters. Peter Jacques, a professor at the University of Central Florida, characterized the mechanisms of environmental denial as having four main components:[14]

- Denying the seriousness of environmental problems and dismissing scientific evidence documenting these problems
- Questioning the importance of policies to protect the environment
- Endorsing anti-regulatory policies
- Portraying environmental protection as threatening Western progress and values

All of these characteristics are part of the basis for denial of climate change in Florida. Conservative think tanks in Florida help promote denial of the reality of climate change. An example is the James Madison Institute, headquartered in Tallahassee. The president of the Institute believes that "the reality is that we don't know why the Earth cools or warms" and that "manmade global warming" is not "settled science." The Institute also published an article in 2008 calling global warming a "cult" and analogizing "red" communism to "green" environmentalism.[15]

Only 34 percent of Americans in May of 2010 believed that "most scientists think global warming is happening."[16] This is a startlingly low number because 97 to 98 percent of the climate scientists most actively publishing support the main conclusions of the IPCC.[17] In this environment of public opinion, most Florida politicians feel safe in ignoring climate change or denying that it occurs at all, even in the face of strong scientific consensus. Jeb Bush, the governor of Florida for 8 years until 2006, claimed to favor "sound science" but expressed doubt as late as 2009 that global warming was occurring at all.[18] Republican Governor Charlie Crist, elected in 2006, was a surprising exception to the political rule. The issue wasn't mentioned in his campaign to succeed Jeb Bush, nor in his inaugural address on January 2, 2007. The day after inauguration, however, St. Petersburg Mayor Rick Baker told Crist that the state's Century Commission for a Sustainable Florida (chaired by Baker) was about to recommend that Florida cut greenhouse gas emissions. On February 6, the governor met with Terry Tamminen, a former aide to California Governor Arnold Schwarzenegger, who was touring the country trying to get governors to act on climate change.

On March 6, Crist included the issue as a priority in his State of the State speech to the legislature. In a speech filled with conventional paragraphs about how Floridians were being "crushed" under the weight of property taxes and how he proposed hiring more "reading coaches" for the education system, he included a surprise:

> It is worth noting that the dramatic rise in our insurance premiums did not occur without cause. It occurred in large part because of an equally dramatic rise in the number and intensity of hurricanes that battered our state in recent years. This simple fact reflects a challenge that we ignore at our own peril. I am persuaded that global climate change is one of the most important issues that we will face this century.

> With almost 1,200 miles of coastline and the majority of our citizens living near that coastline, Florida is more vulnerable to rising ocean levels and violent weather patterns than any other state. Yet, we have done little to understand and address the root causes of this problem, or frankly, even acknowledge that the problem exists.

> No longer. Following this legislative session, I will bring together the brightest minds to begin working on a plan for Florida to explore groundbreaking technologies and strategies that will place our state at the forefront of a growing world-wide movement to reduce greenhouse gases. Florida will provide not only the policy and technological advances, but the moral leadership, to allow us to overcome this monumental challenge.[19]

Crist followed up. In the summer of 2007 he held the first of two annual climate change "summits." At the first summit he signed three significant executive orders. One of them spelled out his numeric targets for reducing greenhouse gas (GHG) emissions:

- by 2017, reduce GHG emissions to 2000 levels;
- by 2025, reduce GHG emissions to 1990 levels; and
- by 2050, reduce GHG emissions to 80 percent below 1990 levels.[20]

He also requested the Florida Building Commission to consider rules mandating a 50 percent increase in building energy efficiency and created the Governor's Action Team on Energy and Climate Change with a report due by October 2008. However, there were no specific water management directives in his executive orders and not much in the final report from the Action Team, either. The Legislature adopted only a fraction of Governor Crist's recommendations and, while running for the U.S. Senate at the end of his term, he scrapped the third of his annual climate change summits.

Water management in Florida plays a significant role in the climate change problem. The nexus among water, energy, and climate is especially important for Florida. Although Florida emits only a very small fraction of global CO_2, it still adds many millions of tons of greenhouse gases to the atmosphere. In combination with other larger sources, Florida's emissions are changing the global climate on which Florida water resources depend. The water–energy connection poses new challenges for water management in Florida.

ENERGY IN THE WATER

Taking all the pieces of the energy–water nexus as a whole, water-related energy use is about 13 percent of all electricity produced in the United States.[21] The U.S. Environmental Protection Agency estimates that about 8 percent of all energy used in the United States is used to pump, transport, treat, and heat water.[22] This percentage is much smaller than in other sectors of the economy, such as transportation. However, many scientific organizations and governments have advocated reducing the world's emissions of greenhouse gases by 2050 to only 20 percent of the levels today (or even of that in 1990). If water-related activities continued to be responsible for the same output of emissions in 2050 as today, they would amount to a major fraction of the total. That is clearly not reasonable or equitable because there will be so many other demands for the available greenhouse gas allocation. The water supply, wastewater treatment, and water end-user communities will have to make their own significant contributions to reducing greenhouse gas emissions if the global goal of large reductions is to be met. The first steps are to be aware of the amount of energy embedded in water production and use.

Obtaining and Conveying Water. Energy is used when water is pumped out of an aquifer, river, or lake. For example, the South Florida Water Management District uses immense quantities of diesel fuel to operate its massive pump stations. Most pumping by other agencies, however, is accomplished by electric pumps, which are energized mostly by the combustion of other fossil fuels. The farther the water has to be pumped for drinking water treatment, the higher the energy demands and the resulting greenhouse gases.

A new trend in Florida water management, which also uses extra energy, is pumping water from the deep Floridan Aquifer rather than from sources closer to the surface. This source has been avoided previously because that aquifer usually has much higher concentrations of dissolved solids. Removing these materials dissolved in the water to make it drinkable is energy intensive.

Treating Water. It takes considerable energy at the drinking water treatment plant to make water safe to drink. As a national average, about 1,400 kWh of electricity is used in the production of a million gallons of drinking water.[23] Florida traditionally has relied on relatively clean sources of groundwater for its public water supplies. Proposed new surface water sources like the St. Johns River will require more energy to treat than groundwater because surface water usually has more dissolved and suspended substances in it, including human pathogens. Developing this kind of alternative water supply will mean higher levels of drinking water treatment (which mean higher energy use and higher cost). Much of this energy-intensive treatment serves no public health purpose because up to half of the drinking water in Florida is used only for irrigating landscapes.

Citizens sometimes write to the Florida Department of Environmental Protection with what they think is a simple answer to demands for more water: desalt the ocean water that nearly surrounds Florida. One reason why this is not practical (except for unusual circumstances) is that desalination costs more than almost any other source of water. The amount of CO_2 emitted in seawater desalination varies with the process and the source of fuel. An Australian study estimates that large desalination plants emit 16,700 to 25,100 pounds of CO_2 for each million gallons of water produced.[24]

Distributing Water. Florida's flatness plays a role here. Unlike many other states, there is not enough topographic relief to store water in reservoirs high up in the hills and then let gravity force it down pipelines to customers. In Florida, water usually has to be pumped every foot of the way. As noted in Chapter 11, Saving the Water, the average Florida home has an amount of drinking water delivered to it each day equal to the weight of a small car. The U.S. Environmental Protection Agency estimates that 3,300 kWh of electricity are associated with each million gallons of surface water delivered to houses. This includes the electricity used in pumping the water from the source, transporting it to the treatment facility, treating it to drinking water standards, and distributing it to homes.[25]

Another growing use of energy in water distribution could be long-distance pipelines, if proposals to ship water from north Florida to central Florida ever became a reality. Long pipelines mean more energy use and more greenhouse gas emissions. In Arizona, pumping water to cities down the route of the Central Arizona Project is the single largest use of electricity in the state.[26]

Using Water. End users often apply more energy in using the water than all of the steps up to delivering it to them. Heating water alone accounts for three-fourths of the energy embedded in water.[27] Some industrial and commercial users also use a great deal of energy to heat and/or circulate water in their production processes.

Most of these water and energy end-use practices could be reduced significantly, and thereby reduce the emission of greenhouse gases.

Collecting and Treating Wastewater. About half of the drinking water delivered in pipes to homes is then routed through different pipes to a wastewater treatment plant, often many miles away. That involves pumping again, plus the energy used in wastewater treatment. (There is energy embedded in the chemicals used in both drinking water and wastewater treatment.) About 1,800 kWh of electricity is used in treating each million gallons of domestic wastewater.[28]

All of these energy expenditures produce greenhouse gases. Municipal wastewater treatment facilities in Florida emitted about 2.23 million tons of greenhouse gases in 2005.[29] That total will increase unless renewable energy sources are used to power the wastewater treatment process.

WATER IN THE ENERGY

The water−energy relationship goes the other way, too. Not only does water use involve energy use, but energy production involves a great amount of water use. Between 3,000 and 6,000 gallons of water are required to light up one 60W incandescent bulb for 12 hours a day for a year. Compact fluorescents require about a third less. About 140 gallons of water are consumed (mostly seawater) for each 1,000 kWh of electricity generated at thermoelectric power plants in Florida.[30]

The amount of water varies with the way that the energy is produced. Of special interest today are proposals to grow biofuel crops in Florida on an immense scale. These are described in the previous chapter (Farming the Water), but it is noted here that by far the greatest amount of water use in making biofuels is at the farm level and not at the production refinery. The Congressional General Accounting Office estimates that the amount of water used in making ethanol from corn varies immensely (from 7 to 321 gallons of water per gallon of ethanol), depending on geographical location and the amount of irrigation. Even for biofuel crops that can be grown without supplemental water, the maximum yield may be produced only with irrigation.[31]

A CHANGING CLIMATE WILL CHANGE FLORIDA WATER MANAGEMENT

Florida's water resources (and the state's 18 million people) may have to make more vigorous and difficult adaptations to the effects of climate change than any other state. Climate change is added on top of how the state's natural systems have already been impaired by more than a century of draining and dredging. A number of reports from Governor Crist's Action Team on Energy and Climate Change, the Florida Oceans Council, Florida Atlantic University, Florida State University, the Florida section of the American Water Works Association, and the Florida

Department of Environmental Protection detail the risks. The main themes can be summarized briefly.

It Is Getting Hotter. Nine of 22 long-term weather stations in Florida recorded the summer of 2010 as the hottest ever.[32] Global temperatures for calendar year 2010 were tied with 2005 for the highest on record. The Northern Hemisphere had the hottest year on record in 2010.[33] The future appears to hold large additional temperature increases. The U.S. Global Change Research Program forecast that the average U.S. temperature in the southeastern United States will increase an average of 4.5 °F under a lower-emission scenario by 2080 and about 9 °F under a high-emission scenario. Summer temperatures would rise an even more oppressive 10.5 °F.[34]

The Oceans Council calls it "probable" that sea-surface temperatures will continue to rise for the next century. Fresh waters are affected by warmer temperatures, too. Dissolved oxygen levels are harder to maintain in warmer water, while biological activity is enhanced. Algal blooms are already much more common in Florida's hot summers.

The Sea Level Is Rising. Approximately 4,500 square miles of Florida are within 4.5 feet of current sea level, and the length of the state's coastline is second only to Alaska's.[35] Not only is the coastline of the state very long, much of it is also very gently sloping. That means that a rising sea can advance very far inland with only a small change in sea level. Therefore, the most obvious climate change hazard for Florida water resources is a rise in the level of the sea. The key question for water management is how high it will go.

From about 1960, the world's oceans have risen about 1.8 centimeters per decade.[36] The forecasted sea level rise has some scientific uncertainty because it depends on the amount of greenhouse gases added to the atmosphere in the future. The IPCC projected in 2007 a global sea level rise of 0.18 to 0.59 meters in the twenty-first century. The panel noted, however, that they did not "assess the likelihood, nor provide a best estimate or an upper bound for sea level rise" and that the "upper values of the ranges are not to be considered upper bounds for sea level rise."[37] Some other projections of future sea level are higher. For example, a committee of scientists advising the Miami-Dade County Climate Change Task Force recommended planning for several feet of sea level rise. They predicted that when the IPCC revises its projection in 2012, the international body would come up with higher estimates than in 2007, "at least in the 3–5 foot range in this century." Moreover, they projected that the rise in sea level would accelerate after 2100.[38]

The rise in sea level has multiple effects. Coastal property, including water facilities, will be at heightened risk of flooding. The more rapid the rise in sea level, the more challenging it will be for water management in a peninsula state. Flood control will become more difficult and costly. The state's coastline will erode more rapidly. Saltwater from the ocean will tend to push freshwater aquifers inland and make coastal wellfields unusable. Researchers at Tufts University estimate that $130 billion of residential property in Florida could be inundated by 2060. A rising sea level also threatens half of the state's recreational beaches, 140 drinking water plants, 2 nuclear power plants, 247 gas stations, and hundreds of other sites that

Figure 14.1 *Sea level rise is an example of the many threats to Florida presented by climate change*

contain hazardous materials.[39] The actual threat depends on how fast the sea rises and how high, about which there is still an unfortunate degree of uncertainty.

The Oceans Are Acidifying; Coral Reefs Are Threatened. Climate change will mean more than hotter air temperatures and rising sea levels. For example, climate change is leading also to the acidification of the oceans. This occurs when atmospheric CO_2 dissolves in ocean water and creates weak acids. By absorbing about 2 gigatons of carbon dioxide a year today (about a fourth of global emissions), the oceans have slowed the increase in greenhouse gases in the atmosphere.

The downside of this chemical reaction, however, is an acidifying effect on oceanic waters. Since the age of industrialization began in the eighteenth century, the acidity of the ocean's surface water has increased by 30 percent (one-tenth of a pH unit). This rate of acidification is a hundred times faster than the oceans have undergone for many millions of years. The changes in ocean acidity will have far-reaching environmental effects for Florida and will occur regardless of how successful the state might be at adapting to changed precipitation patterns or higher sea levels. In combination with higher sea temperatures, acidification may damage or even eliminate the state's subtropical coral reefs. Coral animals will be less able to capture the calcium in ocean water as it becomes more acidic.[40]

Rising sea water temperature is also a threat because the oceans adjoining Florida are already 1.0 to 1.5 °C warmer than a century ago. Heat stress can lead to mass coral bleaching. Coral reefs are limestone structures built by corals (animals related to anemones). Corals share their structures with a type of colorful algae from which the coral animals get most of their food energy. In coral bleaching episodes, the algae die or are expelled from the coral, causing the corals to become lighter in color, potentially starve to death, or transform into almost pure white stone. Heat stress can kill corals directly. The Florida Oceans Council called it "probable" that the "thermal tolerance limits of some coral species will be surpassed" as the oceans warm even more.

Rainfall Patterns Are Changing. It was not by accident that Florida agencies have had names like the Everglades *Drainage* District and the Central and Southern Florida *Flood* Control District. Higher air temperatures will mean, on average around the globe, more rain. However, existing climate models do not do as good a job of forecasting rain patterns for Florida as some other locations. There may be either more rain or less rain for the southeastern United States and the Florida peninsula. It is expected, however, that the variability of rainfall will increase, meaning more Florida extremes: more floods and more droughts. Even small additions to rainfall can cause surprisingly large flood damage. An increase of only 1 inch of rain a year in Florida coastal counties results in a 19.4 percent increase in the likelihood of a flood. Increasing annual rainfall by 4 inches increases the annual chances of experiencing a costly flood by 111 percent.[41]

If droughts become more common in the southeast, the water supply challenges will be exacerbated. Increasing population with decreasing rainfall is a difficult combination for water suppliers to face. It will become harder to achieve desired flows in the Apalachicola River, which will make it more difficult for Florida to sign or maintain a meaningful interstate agreement with Georgia on the management of that watershed.

Hurricanes and Other Tropical Storms Are Changing. Hurricanes get their ferocious energy from the warm water of the tropics when it is heated during the long equatorial summer. The destructive power of Atlantic Ocean hurricanes has already increased in the last 40 years.[42] Hurricanes are likely to become more damaging, according to the U.S. Environmental Protection Agency, with higher peak winds and heavier rainfalls.[43]

Global warming could increase the number of tropical storms or make those with the highest velocity winds more common. The IPCC concluded in 2007 that it is more likely than not that hurricanes would become more intense. So did the U.S. Climate Change Sciences Program in 2008. However, the 2009 report from the Florida Oceans Council thought it probable that tropical storms would become less common and only "possible" that severe hurricanes would become more common. It is interesting that the 2010 hurricane season tied with 1969 for having the second highest number of hurricanes on record. Fortunately for the United States, few reached the mainland.[44] What happens with tropical storms is immensely important for Florida because it already endures more hurricanes than any other state. Higher-velocity winds have much more destructive energy than those in lesser hurricanes. If climate change intensifies hurricanes or makes the most damaging ones more common, the results could be catastrophic for human lives and property in Florida. Even if climate change does not change hurricane frequency or intensity, Florida will be at more risk from hurricanes because sea levels are rising. Hurricanes often induce a temporary surge in sea level, elevated by low atmospheric pressure and pushed ahead by their high winds. These effects force sea water levels higher and farther inland. If a hurricane starts from a higher level of the ocean, the surge will be higher, too. Lesser storms can create damages equal to that of larger storms without the effect of sea level rise. For example, researchers at the Florida State University Beaches and Shores Center determined that storms like Hurricane Dennis, which struck Wakulla County, are likely to occur about every 30 years. If the IPCC projection of sea level rise occurs, however, they calculate that a similar storm will do the same level of coastal damage about every 14 years.[45] Similarly, a storm like Hurricane Wilma, which hit Miami-Dade in 2005, would inflict at least 40 percent more economic damage in 2080 (in real terms), merely due to higher sea levels.

Human Health Is at Risk. Making Florida hotter poses health risks due to waterborne diseases. The type of illnesses associated with tropical climates may become more common in a warming region because the vectors of those diseases find the weather more compatible with their life history. Increased rainfall may prove also to increase human disease rates in Florida. Flood events can contaminate the water in private wells or even inundate drinking water treatment plants. Two-thirds of waterborne disease outbreaks in the United States come after high-rainfall events.[46]

Natural Systems Are Threatened. Humans have never attempted a global climate change experiment before, so it is difficult to foresee all of the environmental changes that are likely to occur. It is clear, however, that most of the natural system alterations are likely to be consequential and adverse. The 2009 report from the U.S. Global Change Research Program predicted that "Ecological thresholds are expected to be crossed" in the southeastern states, which would "cause major disruptions to ecosystems and to the benefits they provide to people."[47] A planetary loss like that would be a permanent biological impoverishment for all humankind. There likely would be many losses specific to Florida, such as the state's coral reefs. The southern Everglades could be mostly inundated by rising sea levels, as well as portions of coastal cities. Other important changes are not so obvious. The Florida Oceans Council calls it "probable" that rising sea levels would eventually "pinch-out" tidal wetlands when their natural landward movement is obstructed by seawalls and other structures. They also pointed to the "possibility" of losing more than half of the state's salt marshes. Climate change also is likely to make Florida an even more welcoming home for new subtropical invasive species.

Florida's sea turtles also are at risk from global climate change. Five species of sea turtles inhabit Florida's waters during some time of the year. Like some other reptiles, the warmer the water and sand environment, the higher the fraction of females in the egg clutches. If the temperature rises too high, the sex ratio imbalance can become too great for the species to maintain itself.[48] The nesting beaches upon which the sea turtles depend are also threatened. The state's beaches are the sites of more sea turtle nesting than any other place in the United States and are used by more loggerhead turtles for nesting than any other place in the world. A sea level rise of even 1 to 2 feet would be enough to cover many of the beaches on which sea turtles nest in Florida. Seawalls currently extend already along

14 to 20 percent of the mileage of Florida's sandy beaches. Unlike in the geologic past, expensive human structures will prevent sandy beaches from being recreated farther landward. Two-thirds of the land in Florida less than a meter above sea level is already developed.[49] The sea turtles will have to adapt to the loss of nesting beaches or disappear in part of their range.

It is impossible to trace through Florida's water system all of the consequences of the changing climate. It is certain, however, that the changes will be quite significant. The rise in sea level induced by greenhouse gases in this century is expected to increase inexorably for at least several more meters in the next centuries unless emissions are soon cut dramatically.[50] Delays in reducing the emission of greenhouse gases make irreversible natural resource damage more likely in Florida and over the whole planet. Because so much is at stake, Florida has to be a leader in the national and global effort to prevent those trends from gaining irreversible momentum.

THE NEED FOR AN ADEQUATE WATER MANAGEMENT RESPONSE

Regardless of international efforts to reduce the emission of greenhouse gases, Florida will experience the effects of global climate change, probably to a more severe degree than most other states. Effective action to adapt to that new reality will require a concerted effort by all water management agencies in Florida, guided by an energetic governor.[51] The Governor's Action Team on Energy and Climate Change produced at least a few water-related recommendations; much more detailed recommendations have been made by the Florida Oceans Council and other organizations. The recommendations below are based primarily on those other reports and have four primary themes:

1. *Face up to the problem.* An effective water management response to climate change requires awakening to the necessity for action, such as:
   - Educate Floridians about climate science and the relationships among climate, water, and energy. This would allow Florida residents and visitors to relate their activities to the climate change–water nexus.[52]
   - Investigate and quantify Florida-specific water management greenhouse gas emissions. How much greenhouse gases in Florida come from water supply activities? From wastewater? From flood control pumping?
   - Undertake risk planning for Florida water facilities. The state and the water management districts should lead an effort to identify which drinking water plants are low enough to be inundated in the next century and which part of the coastline might be hammered by this century's hurricanes. After the identification of the risks, appropriate actions are needed to minimize the vulnerabilities.
   - Implement a well-funded research program on climate change and water in Florida. Global climate change models are not yet fine-grained enough

for Florida. The state should help refine current models to better understand climate change effects for the state.

2. *Integrate climate change into water management plans and programs.* Florida water management has to adapt to the new climate change challenge, as it has done for other water challenges in the past.

   o Put climate change at the center of water planning. The threat of climate change for water management should be a foundation of a reinvigorated Florida Water Plan, as well as the strategic plans and regional water supply plans of the water management districts. The state Water Resource Implementation Rule (Chapter 62–40, F.A.C.) could help set priorities in this effort for the Department of Environmental Protection and the water management districts.

   o Align funding with climate change priorities. Reorganize funding programs, like the State Revolving Fund and the Water Protection and Sustainability Program, to emphasize water and energy efficiency as well as the need for adaptation to climate change.

   o Discourage permitting for activities that result in the emission of large amounts of greenhouse gases. The obvious example is seawater desalination, which requires a great deal of energy.

3. *Emit Less Carbon.* Water in Florida, like other activities, needs to be put on a low-carbon diet. The oil that boiled out of the bottom of the Gulf of Mexico in BP's Deepwater Horizon well in 2010 generated enormous concern about Florida's sugar-sand beaches and coastal ecosystems, but the spill represented less than one-tenth of 1 percent of world oil production of 84 million barrels a day.[53] Florida can be a leader in the necessary move away from carbon-based energy. Some of the actions that could be taken to reduce carbon emissions in water management are:

   o Change the rules for financial assistance from the state for water projects. Low-carbon projects should become high-priority investments for state funding.

   o Change the law so that water management activities must consider climate change in their operations and move to reduce greenhouse gas emissions. The 2009 legislature required local governments to consider carbon emissions in their comprehensive plans but placed no similar requirement on water management agencies.

   o Strengthen, rather than eliminate Florida's Energy and Climate Commission. Rick Scott, Florida's new governor, proposed in 2011 a repeal of the 2008 law establishing a nine member public body to address climate change issues – the Energy and Climate Commission (Section 373.6015, F.S.). [54] Moving the important tasks of the Commission to an already-overburdened Department of Environmental Protection will make the related issues less visible and less likely to see action.

   o Take advantage of water management programs that can create sinks for taking up carbon. Researchers at the University of Florida identified "ample opportunities to sequester carbon" in Florida soils.[55] There are ways to manage both natural and developed land to enhance the rate at which carbon in the atmosphere is captured and stored permanently in the state's organic soils.

   ○ Minimize new projects that pump large quantities of water or over long distances. An exception might be made for projects that derive their energy from renewable and low-carbon sources.

4. *Use Less Water.* Turning down the flow from the faucet is like turning down the thermostat. Saving water saves energy and saving energy saves water. The demand for both is growing in Florida, despite very large efficiency opportunities. The more energy we save by becoming more water efficient, the smaller is the state's carbon footprint. Governor Crist's Action Team on Energy and Climate Change called for "intense conservation of all water uses." Mary Ann Dickinson, the president and CEO of the national Alliance for Water Efficiency, explains the energy–water connection very clearly:

> Supplying, treating, and distributing water is an energy-intensive activity. Thus, water efficiency means energy efficiency. Every drop of water saved reduces the need for treatment, pumping and wastewater disposal, resulting in kilowatt/hours saved. Every drop of water not heated or pumped saves fuel, resulting in therms saved. Thus, water efficiency reduces greenhouse gas emissions from those energy sources.[56]

Water conservation is the single best way to reduce water-related energy costs and greenhouse gas emissions.

A NEW CLIMATE FOR WATER MANAGEMENT?

State government failed Floridians in the 1960s because the foundations of its democracy had been subverted. As explained in Chapter 1, Origins, Florida's legislature was the most malapportioned in the country until the United States Supreme Court required all states to use the "one person, one vote" principle. That 1962 decision to require fair apportionment and the subsequent election of new legislators finally reformed the legislature in the late 1960s. A reformed legislature was able to make a fundamental change in Florida water law.

An even more significant malapportionment problem exists today: future generations, who have the most to lose from climate change, have no vote on today's climate change and water management policies. For Florida, there is no conceivable adaptation to the expected rise of sea level in future centuries if we continue to load up the atmosphere with greenhouse gases. Nor is there any way to de-acidify the ocean once the change occurs. We must prevent the impacts rather than figure out how to live with them. Preventing fundamental harm to the state's water resources will require the present generation to take a very long view of water management. It will require even more political and legal determination than was necessary to create the Water Resources Act of 1972.

A watery Paradise is worth it.

NOTES

1. Fernald, Edward A., and Donald J. Patton, eds. (1984) *Water Resources Atlas of Florida*, Florida State University, Gainesville, FL, p267

2. John Holdren is quoted in Thomas Friedman (2008), *Hot, Flat, and Crowded*, Farrar, Straus, and Giroux, New York, p125

3. Archer, David. (2009) *The Long Thaw*, Princeton University Press, Princeton, NJ, p7

4. Ramanathan, V., and Y. Feng. (2008) "On avoiding dangerous anthropogenic interference with the climate system: Formidable challenges ahead," *Proceedings of the National Academy of Sciences*, vol. 105, no. 38, pp14245–14250

5. Atmospheric CO_2 level average concentration during 2010 at the Mauna Loa Observatory was 389.78 ppm: Dr. Pieter Tans, Trends in Atmospheric Carbon Dioxide, NOAA/ESRL (www.esrl.noaa.gov/gmd/ccgg/trends/) Accessed February 2, 2011. Higher CO_2 level than any other time in the last 800,000 years: Allison, I., et al. (2009) *The Copenhagen Diagnosis: Updating the World on the Latest Climate Science*, University of New South Wales Climate Change Research Centre, Sydney, Australia

6. American Association for the Advancement of Science. (2006) *Board Statement on Climate Change*, December 9, 2006

7. National Academy of Science. (2007) "Joint science academies' statement on growth and responsibility: Sustainability, energy efficiency and climate protection," www.nationalacademies.org/includes/G8Statement_Energy_07_May.pdf, accessed 11 July 2010

8. American Association for the Advancement of Science. (2009) Letter from eighteen scientific organizations to U.S. Senate, October 21, 2009, www.aaas.org/news/releases/2009/media/1021climate_letter.pdf, accessed 18 July 2010

9. Revelle, R., and H. E. Suess. (1957) "Carbon dioxide exchange between atmosphere and ocean and the question of an increase of atmospheric CO_2 during the past decades," *Tellus*, vol. 9, pp18–27

10. Archer, op. cit., note 3, p123

11. The National Academies Committee on Stabilization Targets for Atmospheric Greenhouse Gas Concentrations. (2010) *Climate Stabilization Targets: Emissions, Concentrations, and Impacts Over Decades to Millennia*, National Research Council, Washington, DC

12. Allison, op. cit., note 5

13. Matus, Ron, and Donna Winchester. (2008) "Public: Faith trumps science," *St. Petersburg Times*, 15 February 2008

14. Jacques, Peter J., Riley E. Dunlap, and Mark Freeman. (2008) "The organization of denial: Conservative think tanks and environmental skepticism," *Environmental Politics*, vol. 17, no. 3, pp349–385

15. McClure, Bob [president/CEO of The James Madison Institute]. (2008) Letter to Editor, *Fortune*, vol. 157, no. 8; Bebber, Robert. (2008) "The cult of global warming," *Journal of the James Madison Institute*, no. 41, pp38–41

16. Yale Project on Climate Change Communication. (2010) *Climate Change in the American Mind: Americans' Global Warming Beliefs and Attitudes in June 2010*

17. Anderegg, William R. L., James W. Prall, Jacob Harold, and Stephen H. Schneider. (2010) "Expert credibility in climate change," *Proceedings of the National Academy of Sciences*, published online June 21, 2010, doi: 10.1073/pnas.1003187107

18. Carlson, Tucker. (2009) "Jeb Bush: The future of the Republican party," *Esquire Magazine*, www.esquire.com/features/jeb-bush-interview-0809#ixzz0KisUu71s&D, accessed 22 November 2009

19. Crist, Charlie (2007) "Florida state of the state address 2007", http://www.stateline.org/live/details/speech?contentId=186712, accessed January 28, 2011

20. Crist, Charlie (2007) Executive Order 07-126, July 13, 2007, http://www.flclimate change.us/documents.cfm, accessed January 28, 2011

21. Griffiths-Sattenspiel, Bevan, and Wendy Wilson. (2009) *The Carbon Footprint of Water*, River Network, Portland, OR

22. Grumbles, Benjamin H. (2007) "Drops to watts: Leveraging the water and energy connection," *Water Efficiency*, July/August, pp12–13

23. Yonkin, Matthew, Katherine Clubine, and Kathleen O'Connor. (2008) "Importance of energy efficiency to the water and wastewater sector", *Clear Waters*, New York Water Environment Association, vol. 38, pp12–13

24. Proust, Katrina, Stephen Dovers, Barney Foran, Barry Newell, Will Steffen, and Patrick Troy. (2007) *Climate, Energy and Water: Accounting for the Links*, Land & Water Australia, Australian National University, Australian Capitol Territory, pp1–68

25. U.S. Environmental Protection Agency, WaterSense Program. (2008) *Water-Efficient Single-Family New Home Specification Supporting Statement, 2008*, pp1–7

26. Megdal, Sharon. (2009) "Cleaning dirty air risks costlier Arizona water," *Arizona Republic*, 1 November 2009

27. Griffiths-Sattenspiel et al., op. cit., note 19

28. U.S. Environmental Protection Agency, op. cit., note 23

29. Governor's Action Team on Energy and Climate Change (2008) *Final Report*, Chapter 2, "Inventory and projections of Florida GHG emissions," October 15, 2008

30. Torcellini, P., N. Long, and R. Judkoff. (2004) "Consumptive water use for U.S. power production," ASHRAE Winter Meeting, January 24–28, 2004, NREL/CP-550-35190, National Renewable Energy Laboratory, Washington, DC, pp1–11

31. Mittal, Anu. (2009) *Energy and Water: Preliminary Observations on the Links between Water and Biofuels and Electricity Production*, General Accounting Office, Washington, DC

32. Natural Resources Defense Council. (2010) *The Worst Summer Ever?*

33. National Oceanic and Atmospheric Administration, National Climatic Data Center. (2011) "State of the Climate: Selected Global Highlights for 2010" http://www.ncdc.noaa.gov/sotc/ accessed February 2, 2010

34. Karl, Thomas J., Jerry M Melillo, and Thomas C. Peterson. (2009) *Global Climate Change Impacts in the United States*, Cambridge University Press, New York

35. Harrington, Julie, and Todd L. Walton, Jr. (2007) *Climate Change in Coastal Areas in Florida: Sea Level Rise Estimation and Economic Analysis to Year 2080*, Center for Economic Forecasting and Analysis, Florida State University, Gainesville, FL

36. Intergovernmental Panel on Climate Change. (2007) *The Physical Science Basis*, Fourth Assessment Report, Working Group I Report, Intergovernmental Panel on Climate Change, Geneva, Switzerland, p387

37. Intergovernmental Panel on Climate Change. (2007) *Climate Change 2007: Synthesis Report*, Fourth Assessment Report, Summary for Policymakers, November 17, 2007, Table SPM.1, Intergovernmental Panel on Climate Change, Geneva, Switzerland

38. Miami-Dade County Climate Change Task Force, Science and Technology Committee. (2008) "Statement of sea level in the coming century" is included within the Miami-Dade County Climate Change Advisory Task Force *Second Report and Initial Recommendations* presented to the Miami-Dade Board of County Commissioners, April 2008

39. Stanton, Elizabeth, and Frank Ackerman. (2007) *Florida and Climate Change: The Costs of Inaction*, Global Development and Environment Institute, Tufts University, Medford, MA

40. Ibid.

41. Brody, Samuel D., Sammy Zahran, Praveen Maghelal, Himanshu Grover, and Wesley E. Highfield. (2007) "The rising cost of floods: Examining the impact of planning and development decisions on property damage in Florida," *Journal of the American Planning Association*, vol. 73, no. 3, pp330–345

42. Karl, op. cit., note 32

43. U.S. Environmental Protection Agency. (2009) *Endangerment and Cause or Contribute Findings for Greenhouse Gases Under Section 202(a) of the Clean Air Act*, Technical Support Document, pES-4

44. National Oceanic and Atmospheric Administration "Extremely Active Atlantic Hurricane Season was a "Gentle Giant" for U.S." November 29, 2010. http://www.noaa news.noaa.gov/stories2010/20101129_hurricaneseason.html, accessed on January 31, 2010

45. Harrington et al., op. cit., note 33

46. Curriero, Frank C., Jonathan A. Patz, Joan B. Rose, and Subhash Lele. (2001) "The association between extreme precipitation and waterborne disease outbreaks in the United States, 1948–1994," *American Journal of Public Health*, vol. 91, no. 8, pp1194–1199

47. Karl et al., op. cit., note 32

48. Wilcove, David S. (2008) *No Way Home: The Decline of the World's Great Animal Migrations*, Island Press, Washington, DC, pp164–165

49. Titus, J. G. et al. (2009) "State and local governments plan for development of most land vulnerable to rising sea level along the US Atlantic coast," *Environmental Research Letters*, vol. 044008

50. Solomona, Susan, Gian-Kasper Plattner, Reto Knutti, and Pierre Friedlingstein. (2009) "Irreversible climate change due to carbon dioxide emissions," *Proceedings of the National Academy of Sciences*, vol. 106, no. 6, pp1704–1709

51. In mid-2009, the author of this book and Jessica Bolson, a scientist from the University of Miami on temporary assignment with the Florida Department of Environmental Protection (FDEP), authored a report for FDEP on climate change, called *Framework for Action: Water Management and Climate Change in Florida*. The recommendations here build on that report.

52. Griffiths-Sattenspiel et al., op. cit., note 19

53. U.S. Energy Information Administration. (2010) "International energy statistics for petroleum," http://tonto.eia.doe.gov/cfapps/ipdbproject/IEDIndex3.cfm?tid = 5&pid = 53& aid = 1, accessed 3 July 2010

54. Governor's Budget Recommendation Conforming Bill, Florida Energy and Climate Commission, 32 pp., February 7, 2011

55. Mulkey, Stephen, Janaki Alavalapati, Alan Hodges, Ann C. Wilkie, and Sabine Grunwald. (2008) *Opportunities for Greenhouse Gas Reduction through Forestry and Agriculture in Florida*, School of Natural Resources and Environment, University of Florida, Gainesville, FL

56. Dickinson, Mary Ann. (2008) "Water conservation through national advocacy," *World Plumbing Review*, issue 2, pp12–13

PART IV

THE WATER FUTURE

CHAPTER FIFTEEN

Will Florida Get the Water Future Right?

M arjory Stoneman Douglas, the renowned Florida environmentalist and author of *Everglades: River of Grass*, was 25 years old when she moved to Florida in 1915. During her long life she saw the damage done by many attempts to remake the watery ecosystems of south Florida. As the founder of Friends of the Everglades in 1969, she led a long fight for restoration. She drew a lesson from that experience. On her hundredth birthday in 1990, she had a ready answer when asked if she was optimistic about the survival of the Everglades: "I am neither an optimist nor a pessimist. I say it's got to be done."[1]

The Florida water management system gets a lot done right. It has a single legal doctrine for water above and below ground. State law designates water as a "public resource" rather than private property. Withdrawal permits in Florida are required for all uses (except use from domestic wells). The Florida regional water management districts have large annual budgets, which allow them to undertake ambitious projects. Unlike most states, the Florida system of water management is organized by hydrologic rather than political boundaries, which allows integration of water management policies across entire watersheds. Since 1972, water has been made available to more than 11 million new Florida residents. Florida managed to make it through recent record droughts without major difficulty. The state water management system (in partnership with the federal government) has nearly completed the Kissimmee River project, which is the largest river restoration project in the world. Billions of dollars are being spent in an attempt to restore the Everglades.

Despite these efforts, many of the state's water problems have not been solved. In part, the water problems of today are the direct consequence of previous successes. Drainage projects, for example, succeeded in lowering the water table enough for farming but also eliminated more than two-fifths of Florida wetlands. Flood control projects deliberately rushed water to the sea to prevent damage to urban areas but also now deprive cities of the water they need in dry times.

Competition for scarce supplies has forced water management districts to cap groundwater withdrawals in large regions. In 2010, the state's water resources are being called upon to support close to 19 million residents. By 2030, population is forecast to increase to 24 million, adding the equivalent of two entire Miami-Dade Counties, with associated demands for more water. There are serious obstacles for the Everglades restoration program. Florida faces climate change problems as serious, or even more challenging, than any other state, which can be addressed adequately only by an effective water law and political system and the level of taxation necessary to support it.

What has to be done for Florida to get the water future right? This book is an attempt to answer that question and includes many specific ideas about promoting sustainability in a state with fragile and vulnerable water resources. A few simple themes connect them. First, avoid doing the wrong thing – actions that make things worse. Second, follow the right advice – accept well-known ideas that deserve full implementation. And third, do the moral thing – water management is inherently a set of ethical choices.[2]

AVOID DOING THE WRONG THING

Douglas was a battler. She called the misguided straightening of the Kissimmee River "stupid beyond words."[3] It might be kinder, and maybe more accurate, to say that civic boosters and water managers of that era had conventional but self-constrained ideas about turning natural resources to human benefit. Misguided ideas must be rethought.

Drop the idea that Florida is running out of water. The persistent idea that a state with average rainfall of 54 inches a year can be running out of water distracts water managers from more important work, forces the construction of unnecessary water supply projects, and costs many hundreds of millions of dollars. The same quest for producing ever more water also energizes the idea of imitating California-style long-distance water pipelines in Florida and an unnecessary state water board bureaucracy. Chapter 11, Saving the Water, sets out a few of the ways in which the same amount of water as is used today can meet growing demands, if used more efficiently.

Let water supply be self-supporting. Tracy Mehan, the former Assistant Administrator for Water of the U.S. Environmental Protection Agency, says, "A subsidized water or wastewater system is not a sustainable one."[4] Unless water users have to pay all of the costs for their water and wastewater, they overuse water – induced by artificially cheap prices. Many millions of dollars sent to water utilities and wastewater utilities by legislators and water management districts to reduce water bills doesn't lower the real cost of water. It only makes taxpayers fund water services rather than water customers.

Cut back on wasteful agricultural subsidies. As discussed in Chapter 13, Farming the Water, agriculture is the largest withdrawer of water (and by far the largest consumer of water) in the state, even though farming contributes a tiny fraction of the state's total domestic product. Huge quantities of water are used to produce low-value agricultural crops.

Florida agriculture should be put on a competitive basis with other businesses that need water. By cutting back on government subsidies and trade barriers, local, state, and federal agencies will save on tax dollars and foregone revenues. Water quality will improve if farmers are not given powerful incentives to grow crops that require massive amounts of fertilizers and pesticides. Lastly, inefficient agricultural irrigation will convert to more efficient practices, which will save water for more productive economic uses.

Keep the governing boards in the permit business. The 2009 legislature took away from the water management districts' governing boards the power to participate in the issuance of water use and environmental resource permits. Only the executive director was to make those decisions. In 2010, the legislature allowed the governing boards to decide whether they wanted to delegate all permits to staff, but still required that all proposed permit denials go to the boards for action. The default for permit decisions should not be issuance. The governing boards that run the districts, rather than their appointed staff, should take the heat for all critical permit decisions.

Set a stronger example of water use efficiency The existing requirements in Florida water law for water use efficiency should be strengthened and then implemented rigorously. Why should automatic landscape irrigation systems be installed in new homes almost as routinely as air conditioning systems? The water conservation solution is not in fine-tuning the hardware of the automatic irrigation system but in not installing the systems in the first place. Amy Vickers, author of the *Handbook of Water Use and Conservation*, explains: "In terms of total water savings, increasing the water efficiency of the typical high-volume residential automatic lawn irrigation system is about as beneficial as increasing the fuel efficiency of a Hummer or a sport utility vehicle."[5]

Much more can be done to promote water conservation in Florida. A first step would be for the Florida legislature to enact a law patterned upon the one adopted in California in 2009 that requires public water supply to reduce per capita use by 20 percent by 2020.[6] Saving water would save both water and energy.

Stop giving the stuff away. As explained in Chapter 12, Charge a Few Nickels for Water, a tiny water use fee would be a very powerful tool for water resource sustainability. A fee provides an indispensable price signal to large water users that wastefulness is expensive. Such a fee is strongly recommended in the *Riparian Model Water Code* of the American Society of Civil Engineers. A very small fee per thousand gallons of water withdrawn could also be a partial substitute for unpopular property taxes currently paid to the water management districts.

There is, in economics, no free lunch, and there should be no free water in a state with water challenges as large as Florida's.

FOLLOW THE RIGHT ADVICE

For the most part, there is little mystery about what should be done to improve water management. Many studies and commissions have addressed the problems of water management in Florida. Prime sources of good advice can be found in the

1971 Governor's Conference on Water Management in South Florida, in the 2008 Water Congress, and the 2004 Riparian Model Water Code from the American Society of Civil Engineers.

1971 Governor's Conference on Water Management in South Florida

Governor Reubin Askew pointed to contemporary water supply and water quality problems, emphasizing that "we're all to blame."[7] He said that the "terrible challenges" in south Florida must be solved because "the alternative will be disastrous." The governor asked the attendees to state their answers "clearly, bluntly, and forcefully." The attendees, guided by John DeGrove and Art Marshall, accepted the governor's dire warnings and were as direct as the governor had requested. Their recommendations went far beyond the simple water supply issues that seemed at one level to be the reason they had gathered:

- There should be no further draining of wetlands for any purpose.
- At the state level, there must be an agency or board that has all the power necessary to develop and ensure implementation of a comprehensive land and water use plan for the State. The agency or board, whichever it may be, should report to the Governor.
- The State must have an enforceable, comprehensive land and water use plan. This plan must be developed immediately. It must be designed to limit increases in population and machines, with their attendant demands on the water supply, to a level that will insure a quality environment.
- Water quality is a far graver problem in the long run than is water quantity.
- We believe in the insistence on the highest treatment known of our effluents.
- Present funding for environmental protection must be greatly enlarged to accomplish the common goal of protecting the economic and environmental values of this State.[8]

The following year, in the 1972 legislative "Year of the Environment," Governor Askew followed up and got the legislature to rewrite the Florida Water Resources Act. Earl Starnes, one of the governor's advisers, emphasized how committed Askew was:

> Reubin Askew was very powerful. He controlled what he wanted to do, and it got done. That was a good situation to work in, particularly because we could have messed this program up so badly without gubernatorial support. It was absolutely essential.[9]

With Askew's leadership, many of the important water supply development recommendations of the conference became state law. Other fundamental recommendations however, like those in the list above, were ignored or not fully implemented. They still merit attention from the governor and legislature.

2008 Water Congress (Sponsored by the Century Commission for a Sustainable Florida)

The 2008 Water Congress, although modeled in part on the 1971 Governor's Conference on Water Management in South Florida, had important differences. Just as in 1971, a severe drought prompted a statewide assembly. Both were based in part on a perception of crisis. Both sets of participants worked in five separate subgroups and then approved joint recommendations in plenary sessions. They had similar numbers of volunteer participants (150 in 1971 and 120 in 2008). Law professors at the University of Florida prepared recommendations for the congress on changes in Florida water law, taking their cue from Professor Frank Maloney's work at the University of Florida that resulted in the 1971 Model Water Code.[10]

However, the Water Congress was a creation of the Century Commission for a Sustainable Florida and not of Governor Charlie Crist. Unlike Reubin Askew in 1971, Crist made no appearance at the 2008 congress. The recommendations from the attendees were directed at the Century Commission rather than to the state's chief executive. The top recommendation was for the state to continue subsidizing the cost of developing alternative water supplies, which caused the *St. Petersburg Times* to editorialize, "Money trumped conservation as the top issue."[11] Other ideas were more visionary, such as setting a per capita goal for water use in Florida and giving water conservation equal access to water supply development funds.

Very few of the recommendations from the Water Congress were enacted the next year, unlike the 1971 Conference. Nonetheless, the Water Congress effort is significant and its 18 recommendations warrant serious consideration. (See Appendix 2 for the full set.)

Update Florida Water Law

Dan Tarlock, principal author of the 2004 report of the Western Water Policy Review Advisory Committee, called an effective state water law regime the "best guarantee that water will be used in an environmentally sustainable manner."[12] Florida's water law has not been fundamentally rethought since its creation in 1972. After the passage of almost four decades and the addition of 10 million residents, it is time for an update that preserves the strong points in the current system.

The most authoritative recommendations on reforming state water law today come from the decade-long effort of the Model Water Code Project of the American Society of Civil Engineers (ASCE).[13] The ASCE authors were able to incorporate concepts of sustainability not available to the authors of the 1971 Model Water Code. According to ASCE, sustainability must "become the pervasive criterion of both public and private water management" and that "in a very real sense, this entire Code is a structure of fostering 'sustainable development.'" Water managers must "limit the use of water to uses that do not permanently impair the biological, physical, and chemical integrity of the resource."[14]

Although the Florida Water Resources Act and some water management districts mission statements call for water resource sustainability, they do not require it nearly as plainly as the ASCE Model Water Code. Florida water law should be updated to reflect the sustainability policies in the ASCE code.

Today's political realities will make it hard to follow this call for statutory sustainability. Twenty-first century Florida has made something of a return to the "Pork Chop Gang" rule that stood in the way of modernization in the 1950s and 1960s. While the old malapportioned legislative districts are no longer legal, they have been replaced with elaborate gerrymandering that makes it difficult to challenge incumbents. From 2000 to 2009, only 10 of 505 incumbent legislators were defeated in a bid for re-election. Only four other states had higher records of re-election.[15] A related political distortion is evident in the split of political parties for seats in the Legislature. Florida voters split almost 50/50 between the Democratic and Republican candidates for Governor in 2010 but one party (the Republicans) holds two-thirds of the seats in both the state House of Representatives and Senate. This system does not accurately represent the will of Floridians, just as the Pork Chop Gang did not.[16]

Like the Pork Choppers, today's elected officials have little reason to fear the electorate and too many politicians use offices for personal benefit. Federal prosecutors convicted more public officials of corruption in Florida from 2000-2009 than in any other state.[17] A statewide grand jury on public corruption in late 2010 reported on what "can only be referred to as Florida's Corruption Tax".[18]

In 2009, Florida's House Speaker, Ray Sansom, was forced to resign for ethical reasons. He had been indicted for grand theft and conspiracy in steering $35 million to a state college that later employed him for $110,000 a year and for a separate $6 million to build an airport hanger at the college that would benefit a friend and campaign contributor. A judge later dismissed the charges because "not every wrongful conduct is a crime." Judge Lewis added, however:

> A fair reading of the grand jury's presentment should give pause to members of the Legislature, and anyone else who cares about public trust and confidence in our government institutions. The grand jurors were extremely critical of the conduct of Mr. Sansom, concluding that he had "violated the trust that the citizens of Florida should expect from its elected representatives." But they were also critical of a process and a culture in the Legislature that not only tolerates such conduct, but seems to think little of it.[19]

Money also dominates Florida politics. Eric Draper, Executive Director of Florida Audubon, concluded at the end of the 2011 Legistative session that "This is a Legislature that is owned lock, stock and barrel by special interests."[20] Rick Scott, the former hospital CEO elected Governor in 2010, spent a record $70 million of his fortune in the political campaign, partly to achieve name recognition in a state where he barely met the seven-year statutory minimum for office. Scott went on to accept millions of dollars from individuals and corporations to put on inauguration festivities. One donor, U.S. Sugar, gave

$200,000 to the Republican Party of Florida and to the inaugural committee, even though Scott had severely criticized the land acquisition deal between the company and the South Florida Water Management District.[21] A U.S. Sugar vice-president and lobbyist got a front-row seat at the inauguration. Not to be outdone by U.S. Sugar, Florida Crystals, its long-time rival in the Everglades Agricultural Area, gave $275,000.[22]

Unbalanced representation and a deep-seated rural–urban conflict are similar dysfunctions to those in Florida before legislative reapportionment in the late 1960s. Legislative leaders in 2010 strongly opposed citizen initiatives to amend Florida's constitution to forbid legislative redistricting from favoring any incumbent or political party.[23] On the ballot in November 2010, 63 percent of voters approved making those policies part of the Florida constitution. However, as the nation redraws its legislative and Congressional districts in 2012, those same lawmakers are charged with implementing the very amendments they opposed. Governor Scott, the designated "Tea Party" gubernatorial candidate, carried none of the six largest counties in the state and relied upon large majorities in rural and suburban areas to overcome a 300,000 vote difference.

The end of Chapter One, The Watery Foundation' quotes long-time Florida politician Bob Graham asking if the state once again was going to be treated as a "commodity" rather than a "treasure."[24] There is evidence that old unsustainable patterns of water mismanagement are returning. For example, the Florida Legislature recently created high obstacles against the adoption of rules necessary to establish reservations of water and minimum flows and levels. They have suspended funding for the very popular Florida Forever land acquisition program that protects water resources. New Governor Rick Scott worked with a willing Legislature to reduce the state's appropriation for Everglades restoration from $50 million to only $20 million. Critically important action on climate change seems very unlikely, in part because Scott doubts that anthropogenic climate change is happening at all.[25]

Although history never repeats itself, there are occasions when it rhymes. Water resource sustainability for Florida may require a political transformation like that which gave rise to a new kind of state legislature in the early1970s.

DOING "RIGHT"

Marjory Stoneman Douglas was a moral leader in Everglades restoration, as Jack Davis, her best biographer, understood.[26] Florida and global sustainability is the ultimate moral test. The scale of human exploitation of the earth's resources is stunning. According to the Millennium Assessment of the United Nations, the amount of nitrogen and phosphorus added to the global material cycle has doubled over preindustrial conditions. Half of all wetlands on the planet have been lost since 1900. The majority of the planet's species are declining in either their numbers or their range. Species are becoming extinct at a rate more than a hundred times greater than the evolutionary average. The Millennium Assessment concluded that "the current demand for many ecosystem services is unsustainable" and that "abrupt and possibly irreversible change may not be widely apparent until it is too late to do much about it."[27]

Climate change is the biggest sustainability challenge. People have put in motion a pace of climate change not experienced during human history. Sea levels are rising at an increasing pace. Changes in climate likely will mean new patterns of damaging floods and drought for Florida. Every year of delay in reducing greenhouse gas emissions makes the climate change problem bigger and more costly to control. Until fossil fuel emissions are controlled, greenhouse gases will continue to build up in the atmosphere, and environmental disasters like the 2010 Deepwater Horizon oil gusher in the Gulf of Mexico will continue to be more than possible. Fortunately, reducing water and energy consumption is often not just the ethical thing to do – it also saves money.[28]

Can Florida's water management system make this transition to a low-carbon society? Large transformations have occurred before. For example, the Southwest Florida Water Management District reconfigured the Four River Basins Project in the 1960s in a more nonstructural direction to reflect new ideas about water. So did the St. Johns River Water Management District in the 1970s for the Upper St. Johns Project. The enactment of the Water Resources Act in 1972 remade Florida water law, reflecting a newly reapportioned and more democratically representative Florida legislature. More recently, the Comprehensive Everglades Restoration Plan (CERP) serves as a foundation for success in the world's largest environmental restoration effort. Water management achievements in the last few decades show a path toward water management success.

Unfortunately, Florida may follow the sustainability path only after the next disaster. It may take a gigantic climate-related crisis before Florida reorients its water management system toward a low-carbon society. The perception of crisis has driven previous water management changes, as after the 1947 floods in southeast Florida and the 1960 floods in west-central Florida. Perhaps a Katrina-style storm, intensified by warmer oceans, will slam into the coastal metropolises of southeast Florida or Tampa Bay and force changes in water policy. Perhaps what will be necessary is the kind of "amazing reform" in the state's political process that came after the demise of the Pork Chop Gang.[29]

Ultimately, getting the water future right must mean doing the moral thing for future generations of Floridians and the water resources that define the state. We must have a "water ethic" to go with Aldo Leopold's famous "land ethic": "A thing is right when it tends to preserve the integrity, stability, and beauty of the biotic community; it is wrong when it tends otherwise."[30] This is ultimately a question of ethics, not of economics or engineering or hydrology. The idea that water exists mostly to make maximum profit, supply the next subdivision, and irrigate the next citrus grove, must give way to a wiser recognition about the water needs of this vulnerable and fragile state.

NOTES

1. Lemoyne, James. (1990) "Everglades sentinel on watch at 100," *New York Times*, 8 April 1990

2. David Feldman's research into the water problems of the southeastern United States led him to conclude that it is essential to include environmental ethics in water policymaking. See his *Water Policy for Sustainable Development* (2007) The Johns Hopkins University Press, Baltimore, MD, p269

3. Douglas, Marjory Stoneman, and John Rothchild. (1990) *Marjory Stoneman Douglas: Voice of the River*, Pineapple Press, Sarasota, FL, p229

4. Mehan, G. Gracy, III. (2009) "The fork in the road: Full-cost pricing or federal subsidies," Water Environmental Federation Conference, October 13, 2009

5. Vickers, Amy. (2007) "Are water managers becoming lawn irrigation managers?" *AWWA Journal*, vol. 99, no. 2, pp87–90

6. California Senate Bill No. 7 (Seventh Extended Session), enacted on November 10, 2009

7. Askew, Reubin O'D. [Florida Governor]. (1971) "Askew, remarks at the Governor's Conference on Water Management in South Florida, Miami Beach, September 22, 1971"

8. Governor's Conference on Water Management in South Florida. (1971) "Statement to Governor Reubin O'D. Askew"

9. Oral History Program. (2000) Interview of Earl Maxwell Starnes, 17 June

10. Klein, Christine A., Mary Jane Angelo, and Richard Hamann. (2009) "Modernizing water law: The example of Florida," *Florida Law Review*, vol. 61, no. 3

11. Pittman, Craig. (2008) "Water congress talking about conservation money," *St. Petersburg Times*, 27 September

12. Tarlock, A. Dan. (2004) "Water law reform in West Virginia: The broader context," *106 West Virginia Law Review*, 495-538, p530

13. American Society of Civil Engineers (ASCE). (2004) *Regulated Riparian Model Water Code*, Section 2R-2-24, ASCE/EWRI 40-003, ASCE, Reston, VA, pix

14. ASCE, op. cit., note 13, pp27–28

15. Silva, Cristina "Politifact: Incumbent Losses in Legislature are as rare as Fair Districts Florida says" *St. Petersburg Times*, July 12, 2010

16. The Florida Division of Election reports that, for the 2010 general election, there were 4,631,068 registered Democrats and 4,039,259 registered Republicans. http://election.dos.state.fl.us/voter-registration/statistics/elections.shtml#2010 Accessed February 3, 2011.

17. U.S. Department of Justice, Public Integrity Section "Report to Congress on the Activities and Operations of the Public Integrity Section for 2009" Washington, D.C. 2010, 59 pp

18. Nineteenth Statewide Grand Jury, First Interim Report, "A Study of Public Corruption in Florida and Recommended Solutions." Case No. SC 09-1910, Ft. Lauderdale, FL, December 17, 2010

19. Leary, Alex "Judge delivers major blow to case against former House Speaker Ray Sansom" *St. Petersburg Times*, October 5, 2009

20. Rockwell, Lilly (2011) "Lawmakers approve controversial permitting bill" *Miami Herald*, May 6, 2011

21. Klas, Mary Ellen "GOP raised millions for Scott, itself" *St. Petersburg Times*, January 18, 2011

22. Klas, Mary Ellen "GOP raised millions for Scott, itself" *St. Petersburg Times*, January 18, 2011

23. Fair Districts Florida, http://www.fairdistrictsflorida.org/our_reforms.php, accessed February 2, 2011

24. Klas, Mary Ellen "Former Sen. Bob Graham warns that oil spill danger hasn't passed" *St. Petersburg Times*, January 14, 2011

25. Pittman, Craig. "Scott, Sink on polar ends of green spectrum" *Miami Herald*, October 25, 2010

26. Davis, Jack E. (2009) *An Everglades Providence: Marjory Stoneman Douglas and the American Environmental Century*, University of Georgia, Athens, GA

27. Hassan, Rashid, Robert Scholes, and Neville Ash. (2005) *Ecosystems and Human Well-Being: Current State and Trends*, vol. I, "The millennium assessment," Island Press, Washington, DC

28. Dernbach, John C., and Donald A. Brown. (2009) "The ethical responsibility to reduce energy consumption," *Hofstra Law Review*, vol. 37, p985

29. Miller, James Nathan "How Florida Threw Out the Pork Chop Gang" National Civic Review, Vol. 60, No. 7, July 1971, pp366-380.

30. Leopold, Aldo. (1966) *A Sand County Almanac*, Ballantine, New York, p262

Milestones of Water Management in Florida

	SFWMD	SWFWMD	SJRWMD	NFWMD	SRWMD	State, Nation, and Beyond
1845						Florida statehood. Federal government grants 500,000 acres of land to the state for "internal improvements."
1850						Wetlands transferred to State based on 1848 Swamp Land Act.
1881	State of Florida sells 4 million acres of land near Lake Okeechobee and in the Kissimmee River basin to Hamilton Disston for 25 cents per acre.					
1883			Lake Beauclair and Lake Apopka connected by canal.			
1888		The first commercial shipment of phosphate is made from the Peace River Valley, where the mineral had been discovered in 1881.				
1894						1894–1895. Great Freeze ends commercial agriculture industry in north Florida.

(continued)

Appendix 1 – *continued*

	SFWMD	SWFWMD	SJRWMD	NFWMD	SRWMD	State, Nation, and Beyond
1901	Audubon Societies propose law prohibiting the killing of birds (except game birds).					
1904	Napoleon Broward elected governor on a promise to drain the Everglades for gardens and farms.					
1906	Formation of the Everglades Drainage District.					
1913	"Drainage of the Florida Everglades is entirely practicable and can be accomplished at a cost which the value of the reclaimed land will justify, the cost being very small." – Florida Everglades Engineering Commission					
1920s	South Florida real estate boom; Carl Fisher transforms a wet, mangrove-fringed island into the resort of Miami Beach.	Saltwater intrusion in St. Petersburg's municipal well fields.				
1926	Major hurricane in southeast Florida, over 200 killed.			Congress authorizes U.S. Army Corps of Engineers to study the Apalachicola River.		
1928	Major hurricane hits Lake Okeechobee, drowning thousands.					
1929	Okeechobee Drainage District formed.					

Year			
1930	Rivers and Harbors Act of 1930 authorized the Herbert Hoover Dike around Lake Okeechobee	Pinellas Water Company begins production from Cosme Wellfield in northwest Hillsborough County	
1932	Beginning of construction of Herbert Hoover Dike around Lake Okeechobee and the Okeechobee Waterway.		
1934	Everglades National Park authorized.		Construction begins on the Cross Florida Barge Canal
1935	"Labor Day Hurricane" hits the Keys, killing 400.		Work suspended on the Cross Florida Barge Canal.
1937	U.S. Army Corps of Engineers completes Herbert Hoover Dike flanking three-quarters of Lake Okeechobee.		
1940		Weekiwachee Springs purchased by St. Petersburg for future water supply.	
1945		Congressional enactment of a 9' × 100' channel in the Apalachicola, Chattahoochee, and Flint Rivers	Citizen's Committee on Florida Water report submitted to legislature.
1947	Regional flooding prompts plan for C&SF Project. Dec. 6. Everglades National Park established	First algal blooms reported in Lake Apopka.	Marjory Stoneman Douglas publishes *River of Grass*. Water Survey and Research Division created within the Board of Conservation.

(continued)

Appendix 1 – *continued*

	SFWMD	SWFWMD	SJRWMD	NFWMD	SRWMD	State, Nation, and Beyond
1948	Congress authorizes the Central & Southern Florida Flood Control Project.					
1949	June 2. Central & Southern Flood Control District established.					
1950		Kissengen Spring ceases to flow.				
1954	Congress authorizes the Kissimmee River Flood Control Project.					Citizen's Water Problem Study Committee formed by Gov. Leroy Collins, reports on March 1, 1955. Final Report from the Florida Water Resources Study Commission, recommends creation of a Department of Water Resources.
1957				Jim Woodruff Lock and Dam on the Apalachicola becomes fully operational, creating Lake Seminole		Ch. 57-380, the Water Resources Act, including the creation of a state Division of Water Resources
1958						Formation by U.S. Study Commission of the Southeast River Basin Study.

Year		
1959	Suwannee River Authority created.	Fidel Castro takes over Cuba, leading to a flood of immigrants to Florida and to an expansion of the state's sugar industry in the Everglades Agricultural Area.
1960	Flooding of west-central Florida from Hurricane Donna and other storms	
1961	Southwest Florida Water Management District established by legislature.	Canal Authority of Florida created, replacing Ship Canal Authority.
	U.S. Army Corps of Engineers recommends the Four River Basins Project be authorized by Congress.	State Board of Conservation is made responsible for supervision of navigation districts.
		Legislature required plugging of flowing artesian wells if water is highly mineralized or unusable
1962	Construction of the Kissimmee Canal begins.	October 23. Congress authorizes the Four River Basins Project with SWFWMD as local sponsor.	Silent Spring by Rachel Carson, published.
			U.S. Supreme Court decision on reapportionment eventually leads to massive changes in state legislature, diminishing the influence of the rural "Pork Choppers."
1963			Ch. 63-336 authorized the creation of "water regulatory districts"
			U.S. Study Commission, Southeast River Basins, submitted Final Report. Florida joined with Alabama and Georgia to create the Southeast Basins Resources Advisory Board.

(continued)

Appendix 1 – *continued*

SFWMD	SWFWMD	SJRWMD	NFWMD	SRWMD	State, Nation, and Beyond
					The Board of Conservation was given general supervision over all regulatory districts.
1964	U.S. Army Corps of Engineers recommends construction of a $12.5 million hurricane levee across Tampa Bay.				Feb. 27. President Johnson presides over groundbreaking ceremony for Cross Florida Barge Canal.
1967					Federal judge requires reapportionment of Florida legislature.
1968 Biscayne National Monument established. Dade County Port Authority holds groundbreaking ceremony for Everglades Jetport.					Congress passes Wild and Scenic River Act. Voters approve a new Florida constitution. Board of Conservation becomes the Department of Natural Resources.
1969	US Geological Survey map shows area in southwestern Polk County as a "caution area" for further water withdrawals.				
1970 Congressional act requiring delivery to Everglades National Park of 315,000 acre-feet per year, or 16.5% of total water deliveries from the C&SF Project, whichever is less.					April 22. First Earth Day

	Jan. 16: Everglades Jetport Pact signed that terminates consideration of an airport site in the Big Cypress.		1970–71. State experiences severe drought.
1971	U.S. Army Corps of Engineers completes the channelization of the Kissimmee River.	Regional Water Supply Development Plan for Escambia, Santa Rosa, Okaloosa, Walton, and Bay counties.	Jan. 19: President Nixon halts work on Cross Florida Barge Canal, leading eventually to the creation of the Marjorie Harris Carr Cross Florida Greenway.
	Disney World opens in Orlando.		Florida legislature requires that all sewage be treated to at least secondary levels.
	Sept. Governor Askew's Conference on Water Management in South Florida.		
1972	First public hearing on the restoration of the Kissimmee River.		Congress passes Clean Water Act.
	Wilson-Grizzle Act of 1972 mandates advanced wastewater treatment for sewage plants discharging into Tampa Bay.		
	SWFWMD issues Order 72-1 placing regulatory levels on the Floridan Aquifer at Cosme-Odessa and Section 21 wellfields.		Environmental Land and Water Management Act; Land Conservation Act, and State Comprehensive Planning Act passed by legislature. Florida Water Resources Act passed, establishing the five water management districts.
			Nov. 7: Florida voters approve the $240 million Environmentally Endangered Lands bond program.
1973	Big Cypress established as first Area of Critical State Concern.	City of St. Petersburg begins pumping from a new wellfield in Pasco County.	Congress passes Endangered Species Act.
		The first governing board meeting is held on November 29, 1973.	

(continued)

Appendix I – *continued*

	SFWMD	SWFWMD	SJRWMD	NFWMD	SRWMD	State, Nation, and Beyond
	Legislature appropriates funds for the Special Project to Prevent the Eutrophication of Lake Okeechobee.				Record flood occurs in the upper reaches of the Suwannee River basin. (1948 remains the record flood on the lower river).	July 1: State system of water management districts established boundaries of the five districts and an interim sixth district.
1974	Big Cypress National Preserve, next to Everglades National Park, was established.	Oct. 25: West Coast Regional Water Supply Authority formed.				Ch. 74-114, authorized creation of regional water supply authorities by interlocal agreement.
1975		Water Use Permitting Program began at SWFWMD.			First SRWMD headquarters purchased for $70,000, converted from the old Suwannee Motel in White Springs.	Legislature adopts Local Government Comprehensive Planning Act.
		May: Court found that the amended SWFWMD boundary meant that SWFWMD would no longer have authority to levy an ad valorem tax without prior voter approval.			Florida DNR Director Harmon Shields suggests piping Suwannee River water to south Florida.	Legislature adopts Aquatic Preserves Act.

	Kissimmee River Restoration Act directs a new Council to study restoration. Central and Southern Florida Flood Control District renamed as SFWMD. Summary Report on the Special Project to Prevent Eutrophication of Lake Okeechobee finds "water delivered to Lake Okeechobee from the Kissimmee River by Canal-38 contributes significantly to the eutrophication of the Lake."	Production begins from Cross Bar wellfield in central Pasco County.	Begins regulation of well construction.	Four bills are filed in the Florida legislature to divert water from north Florida to central and southern Florida.	Legislature creates Department of Environmental Protection and gives it responsibility for general supervision of the water management districts. March 9: Voters pass constitutional amendment authorizing WMD property taxes.
1977	Upper St. Johns River Basin Restoration Project begins.		A consultant to the West Coast Regional Water Supply Authority calls it feasible to pipe water from the Suwannee River.		

(continued)

Appendix I – *continued*

	SFWMD	SWFWMD	SJRWMD	NFWMD	SRWMD	State, Nation, and Beyond
			The U.S. Army Corps of Engineers recommends against resumption of construction on Cross Florida Barge Canal.			
1978						Drafts of State Water Use Plan, based on chapters prepared by the water management districts.
1979	EPA designates the Biscayne Aquifer as a Sole Source Aquifer.	Legislature designates Green Swamp as an Area of Critical State Concern.		Adopted rule for Management and Storage of Surface Waters	The state Environmental Regulation Commission designates the Suwannee an Outstanding Florida Water (OFW).	Draft State Water Use Plan deferred in favor of first developing a state Water Policy Rule (Ch. 17-40, F.A.C., later transferred to Ch. 62-40). Legislature adopts the Conservation and Recreation Lands Program.
1980				Floridan aquifer levels in Ft. Walton Beach area have declined as much as 100 feet below sea level.	The U.S. Army Corps of Engineers issues its Four River Basins Water Management Study which includes proposed pipeline routes from the Suwannee River to the Tampa area.	

1981		Governor Graham establishes the Suwannee River Resource Planning and Management Committee.	SJRWMD issues water shortage declaration.	The District moves to Live Oak. The Resource Planning and Management Committee completes the Suwannee River Management Plan. Gov. Bob Graham, after a canoe trip on the Suwannee, proposes an increase in the documentary stamp tax, which eventually became the Save Our Rivers Trust Fund.	Governor Graham appoints the second Environmental Lands Management Study Committee. State Water Policy Rule (Ch. 17-40, F.A.C.) adopted. Legislature adopts Save Our Rivers Act and Save Our Coast Act. Several winter freezes from 1981–1983 cause much of the state's citrus groves to be relocated farther south.
1982	High water threatens deer herd in Conservation Areas.	Tampa Bay Area Scientific Information Symposium (BASIS I)		Begins consumptive use permitting	Adopts Chapter 40B-2, FAC, Permitting of Water Use. State groundwater rule adopted.

(continued)

Appendix 1 – *continued*

	SFWMD	SWFWMD	SJRWMD	NFWMD	SRWMD	State, Nation, and Beyond
					Suwannee River Floodplain Mapping and Modeling Project begins.	
1983	August 9: Governor Bob Graham announces the Save Our Everglades program.	Tampa Bay Study Committee and the Agency on Bay Management established.	Begins permitting water use district-wide.	Lake Jackson Regional Stormwater Facility completed.	Acquires the 9,315-acre Brunswick Tract in the Lower Suwannee Basin; the District's first Save Our Rivers purchase.	Florida Water Quality Assurance Act with trust fund for water quality improvement.
	August 19: Coordinating Council on Kissimmee Restoration supports "partial backfilling" alternative.					
1984				Began regulation of Agricultural and Forestry Surface Water Mgt.	Highest river stages in the SRWMD since 1973 were recorded in March and April.	Legislature adopts the Warren S. Henderson Wetlands Protection Act. Legislation is also passed creating the Management and Storage of Surface Waters (MSSW) Program under the WMDs, giving them authority to regulate and control stormwater discharges from development.

Year					
1985	Lake Okeechobee Technical Advisory Committee formed.	Adopted first Water Shortage Plan.	SRWMD passes its Works of the District Rule, which requires a 75-foot buffer on major rivers.	Governing board adopted the Works of the District Rule, Chapter 40B-4, FAC.	Legislature adopts the Growth Management Act and State Comprehensive Plan. Florida Rivers Study Committee recommends statutory program for Florida Resource Rivers.
1986					Legislature establishes the nation's first program to clean up contamination from leaking underground petroleum storage tanks. State Water Use Plan adopted by the Department of Environmental Regulation. Legislature enacts the Surface Water Improvement and Management Act.
1987	Everglades designated a Wetland of International Importance.	Grizzle-Figg legislation strengthens and expands the Wilson-Grizzle Act of 1972, mandating advanced wastewater treatment for Tampa Bay.			
1988	Oct. 11: U.S. Attorney sues SFWMD and FDEP for violating federal water quality standards in the Everglades.	SJRWMD begins restoration of Lake Apopka.	SWIM plans written for Alligator Lake, Santa Fe River, Steinhatchee River, Lower Suwannee River, Upper Suwannee River and Falling Creek.		Department of Environmental Regulation added reuse provisions (including mandatory reuse in Water Resource Caution Areas) to the Water Policy Rule.

(continued)

Appendix I – *continued*

	SFWMD	SWFWMD	SJRWMD	NFWMD	SRWMD	State, Nation, and Beyond
1989		Revisions to water use permitting rule, not allowing unacceptable impacts. District declares northern Tampa Bay, eastern Tampa Bay, and Highlands Ridge as Water Use Caution Areas.	SJRWMD issues Water Shortage declaration.	Water Resource Caution Area established for coastal areas of Santa Rosa, Okaloosa, and Walton counties.	Gov. Bob Martinez, after a canoe trip on the Suwannee, appoints the Suwannee River Task Force to conduct a study of the river.	Legislature directs the water management districts to assume most stormwater management responsibilities.
1990	Sugarcane industry is about 450,000 acres. In Water Resources Development Act of 1990, Congress directs a study of environmental restoration of the Kissimmee River.	National Estuary Program established in Tampa Bay.	Florida enacted the Indian River Lagoon Act, which encourages reuse in this area of the central east coast of Florida.		DNR closes large areas of Suwannee Sound to commercial oystering due to high bacteria levels. SRWMD develops the Floodplain Lot Acquisition Program.	Preservation 2000 provides $300 million per year over 10 years to purchase ecologically valuable lands. Cross Florida Barge Canal deauthorized. Legislature passes Growth Management Act, which requires "concurrency" in the delivery of public facilities, such as schools, transportation, and water supply.

1991	May 7: Florida legislature passes the Marjory Stoneman Douglas Everglades Protection Act. May 20: Governor Chiles appears in court and "surrenders" in the Everglades water quality standards lawsuit.	SJRWMD adopts the districtwide water conservation rule.	The Florida Dept. of Health and Rehabilitative Services proposes to impose additional constraints on placement of septic tank drainfields in flood-prone areas.	FDEP publishes guidelines for preparation of reuse feasibility studies. Final Report from the Bluebelt Commission.
1992	March: District judge accepts Settlement Agreement requiring Florida to construct 32,000 acres of stormwater treatment areas and reduce phosphorus pollution in Everglades National Park. August: Hurricane Andrew hits south Florida, causing $16 billion in damages. Water Resources Development Act of 1992 authorizes Kissimmee River Restoration and South Florida Restudy.	SWFWMD combines its three Water Use Caution Areas to establish the Southern Water Use Caution Area. SWFWMD's Water Supply Needs and Sources Report 1992. First comprehensive, districtwide assessment of water supply.	The three states and the U.S. Army Corps of Engineers agree to a comprehensive study of the ACF Basin. Water Resource Caution Area established for Telogia Creek in Gadsden County.	SRWMD implements a small-lot acquisition program. U.N. Conference to increase biodiversity and halt global warming (Agenda 21). Reuse Coordinating Committee established.

(continued)

Appendix 1 – *continued*

	SFWMD	SWFWMD	SJRWMD	NFWMD	SRWMD	State, Nation, and Beyond
1993	July 13: Secretary of Interior Babbit announces Statement of Principles to settle Everglades lawsuits.			28,009-acre tract in Tate's Hell State Forest purchased.	The "Storm of the Century" hits the Big Bend Coastal Area in March.	The Department of Natural Resources and Department of Environmental Regulation are merged into the Department of Environmental Protection.
1994	March 3: Formation of Governor's Commission for a Sustainable South Florida.			Hydrologic restoration of Tates Hell Swamp initiated.	District commits to research with the U.S. Geological Survey to generate information	Legislature creates Water Management District Review Commission.
	May 3: Enactment of Everglades Forever Act.				needed to develop Minimum Flows and Levels for the lower Suwannee River and estuary.	Legislature combines permitting for stormwater and for dredge and fill in the Environmental Resource Permitting program.
						WMDs complete first District Water Management Plans.
1995	Farm Bill allots $200 million for Everglades land acquisition.				Florida and Georgia form the informal Suwannee Basin Interagency Alliance.	Intergovernmental Panel on Climate Change reports again that global warming is occurring.
						Florida Water Plan adopted by the Department of Environmental Protection and District Water Management Plans by the WMDs.

Year						
1996	In the Water Resources Development Act, Congress directs a study of "restoring, preserving, and protecting the South Florida Ecosystem." Florida voters reject a penny-per-pound tax on sugar to pay for Everglades cleanup.				Dairy Cost Share Program begins.	Florida legislature passes Bluebelt Act to protect recharge areas. Governor's Executive Order 96-297 requires the water management districts to submit priority lists and schedules for establishment of minimum flows and levels.
1997	Construction begins on the first Stormwater Treatment Areas.	Mulberry Phosphates mine waste spills into the Alafia River.	River Summit held for the Lower St. Johns River.	Congress passed, and the three states ratified, the ACF Compact. 28,954 acres adjacent to Econfina Creek purchased.	District purchases Atsena Otie Key, the first acquisition specifically to protect coastal resources.	Legislature amends Ch. 373, adopting "local sources first" policies and requiring action on minimum flows and levels and on regional water supply planning. 38% of flow from Florida's domestic wastewater treatment plants is reused.
1998	Tampa Bay Water formed as successor to West Coast Partnership Agreement between District, Tampa Bay Water, Pinellas, Hillsborough, Pasco counties, Cities of St. Pete, Tampa, New Port Richey.			The SRWMD's water quality monitoring network detects significant increasing concentrations of nitrate in the middle Suwannee, mainly from groundwater inflows.		Formation of Cross Florida Greenway.

(continued)

Appendix 1 – *continued*

	SFWMD	SWFWMD	SJRWMD	NFWMD	SRWMD	State, Nation, and Beyond
					USFWS proposes to completely remove the Okefenokee sill.	May 4: Marjory Stoneman Douglas dies, age 108.
1999	March 31. Kissimmee Canal (C-38) backfilling begins.				24 entities form the Suwannee River Partnership to provide technical assistance and funding to implement Best Management Practices for farms.	Gov. Bush forms the Florida Springs Task Force after taking a canoe trip on the Ichetucknee.
	Comprehensive Everglades Restoration Program recommendations presented to Congress.					Florida Forever Act provides $300 million per year for 10 years for land acquisition, water resources protection and supply; ecosystem restoration, and urban parks and open space.
2000	Dec. 11: President Clinton signs the Comprehensive Everglades Restoration Program funding authorized by Congress in WRDA.			Governing board declares a Water Shortage Warning for the entire district.	Beginning in the latter part of 1998, a deficit of rainfall creates the most severe drought in the Suwannee Basin since the mid-1950s; some first-magnitude springs cease flowing.	Growth Management Study Commission formed.

Year			
	Everglades Restoration and Lake Okeechobee Protection Acts passed by Florida legislature.	Northern portion of Lake Jackson drains through Lime Sink.	Gov. Bush and FDEP create the Florida Springs Initiative to implement the recommendations of the Springs Task Force.
2001	SJRWMD issues a water shortage declaration for part of the district.	Regional Water Supply Plan for Region II (Santa Rosa, Okaloosa, and Walton counties) approved	
2002	Florida Bay undergoes a 700-square-mile mass of "black water" caused by an algal bloom. The SFWMD Board votes to spend $120 million to buy out 1,400 landowners on 6,000 acres in the East Everglades. President George W. Bush and Gov. Jeb Bush sign agreement to reserve water under state law for natural system restoration as a condition for receiving federal financial assistance for CERP projects.	SRWMD and FDEP propose the creation of the Suwannee Wilderness Trail.	DEP and other organizations complete the report of the Water Conservation Initiative.
2003	State donates 42,000 acres to expand Everglades National Park	Governors of Florida, Georgia, and Alabama are unable to reach an ACF agreement; Compact is allowed to expire in August.	The US Fish and Wildlife Service designates the Suwannee and portions of the Withlacoochee and Santa Fe rivers as "critical habitat" for Gulf sturgeon.

(continued)

Appendix 1 – *continued*

	SFWMD	SWFWMD	SJRWMD	NFWMD	SRWMD	State, Nation, and Beyond
	Legislature amends Everglades Forever Act changing phosphorous pollution standards, and appropriates $225 million for CERP.					
2004	Beginning of Acceler8 to accelerate Everglades restoration.		SJRWMD releases report on the potential for using the St. Johns River as a water supply source.			Hurricanes Charley, Francis, Ivan, and Jeanne hit Florida in a six-week period.
	State adoption of 10 ppb water quality criterion for phosphorus in the Everglades.					
	First phase of Kissimmee restoration completed.					
2005				Adopts new rules for the permitting of bottled water.	The district adopts minimum flows and levels for Madison Blue Spring.	Conserve Florida program for improving efficiency in public water supply created by Legislature.
						Fourteen hurricanes form to make 2005 hurricane season the busiest on record.
						Legislature creates the Century Commission for a Sustainable Florida.
						Legislature enacts the Water Protection and Sustainability Program to assist in developing alternative water supplies.

Year						
2006	SFWMD and FDEP purchase the last land needed for Kissimmee River Restoration Project. National Academy of Science review of Everglades restoration. Completion of 41,000 acres of Stormwater Treatment Areas.	SWUCA Recovery Strategy approved by SWFWMD governing board.	River Accord: Partnership of the city of Jacksonville, the JEA utility, and FDEP, to improve the heath of the St. Johns River lower basin.	Governing board adopts a reservation on the water resources of the Apalachicola River (including the Chipola River).		The SJRWMD, SFWMD, and SWFWMD begin development of a plan to address water supply concerns in the Central Florida Coordination Area.
2007	Water Resources Development Act passed by Congress and authorizes individual projects in Everglades Restoration Lake Okeechobee reaches record low. Governing board adopts the Lower East Coast Regional Water Availability Rule.		District commences a multiyear $3.5 million St. Johns River Water Supply Impact Study	Phase I (stormwater) ERP in effect. Water Shortage warning issued for entire district. Regional Water Supply Plan for Region V (Gulf and Franklin counties)	The district adopts minimum flows and levels for the Lower Suwannee River.	Legislature reduces (and later eliminates) funding for the Water Protection and Sustainability Program. WMDs reduce property tax millage rates to comply with legislative requirements. Statewide drought.
2008	June: Gov. Crist announces $1.7 billion plan to buy U.S. Sugar, including land in the EAA, mill, and railroad. Nov: Crist announces revised $1.3 billion plan to buy only U.S. Sugar land.	Seawater desalination plant completes construction with $90 million in assistance from SWFWMD.		The Ichetucknee Partnership is formed.		Drought eases in most of Florida except SWFWMD region. August 19: Tropical storm Fay made the first of four landfalls in Florida, releasing record amounts of rain and easing drought.

(continued)

Appendix 1 – continued

	SFWMD	SWFWMD	SJRWMD	NFWMD	SRWMD	State, Nation, and Beyond
						Century Commission sponsors the Water Congress meeting in Orlando.
2009	May 13: SFWMD governing board votes to acquire 73,000 acres of sugar land in the EAA at a cost of $536 million. Picayune Strand, C-111, and Tamiami Trail projects commence construction.	Most of SWFWMD is in Phase III or IV water shortage restrictions throughout this year.	Nov. 1: District rule permanently limits winter irrigation to one day per week (the most stringent in Florida).	Federal judge Paul Magnuson gives Florida a major legal victory in the ACF litigation.	SRWMD and SJRWMD undertake a joint planning effort for the Upper Santa Fe Basin.	Enactment of Senate Bill 2080, which transfers permit decisions from the governing boards to the executive directors. State receives more than $250 million in recession stimulus funding from the federal government, much of it for water supply and wastewater projects.
2010	US EPA requires the construction of an additional 42,000 acres of Stormwater Treatment Areas. August 12: SFWMD governing board votes to acquire 27,000 acres of the U.S. Sugar holdings.	Jan. 3– 13: SWFWMD rules allow almost 1 billion gallons per day to be pumped for agricultural freeze protection during a cold spell, inducing a number of sinkholes and the temporary drying up of over 750 wells.	National Research Council releases third interim report on St. Johns River Water Supply Impact Study	Full Environmental Resource Permitting was implemented by the District and the FDEP on November 1.	SRWMD calls itself a "victim of groundwater pumping."	Congress fails to enact energy and climate bills.
2011						Florida Legislature cuts WMD tax levels, sets their budgets for the first time, makes it harder for citizens to challenge WMD permits.

2008 Water Congress Recommendations

1. Amend, as necessary, any statute, rule or policy so that quantifiable water conservation best management practices are considered an "alternative water supply" and are equally as eligible for funding as capital facility expansion proposals.
2. Achieve dramatic improvements in landscape irrigation efficiency by requiring use of the recommendations found in the report *Landscape Irrigation and Florida-Friendly Design Standards* (where applicable) as a condition of:
    o   Consumptive Use Permits issued by Water Management Districts
    o   Development orders issued by local governments
    o   Development orders for Developments of Regional Impact as reviewed by Regional Planning Councils
    o   Land use amendments reviewed by the Department of Community Affairs
    o   Changes to land development regulations
    o   Environmental Resource Permits issued by the Florida Department of Environmental Protection
3. Set a per capita target or goal for water use and quantifiable best management water practices and provide a stable funding base for the Conserve Florida program directed by Sect. 373.227, F.S., including the statewide water conservation clearinghouse for public water supply.
4. Coordinate public information efforts statewide regarding water conservation, water quality, sustainability, and energy using the most effective methods of conveying the messages and measuring the efficacy of the public information campaigns. Examples include but are not limited to:
    o   Landscape water conservation must be underscored by science-based, practical, and easily understood recommendations for homeowners and do-it-yourself gardeners to follow.

       o Establish public and stakeholder outreach program regarding costs, environmental advantages and effectiveness of water saving technologies.

       o Educate all sectors regarding the interdependency of upland and wetland systems.

5. Amend Florida law to prohibit neighborhood organization and local ordinances from restricting the use of Florida-Friendly landscaping.

6. Support regional partnerships, incentives, and cooperative approaches to addressing long-term water sustainability for Florida. The Water Management Districts, the FDEP, and local governments should aggressively identify opportunities and develop incentives for establishing multijurisdictional utility arrangements or water supply authorities and engage in other areas where such partnerships should be beneficial and cost effective to the public.

7. Support the development of robust incentive-based cooperative funding programs by the Water Management Districts to encourage the development of alternative water supplies and implementation of conservation measures, including the maximum use of reclaimed water that will require seeking state and federal funding to complement water management district funding initiatives.

8. Amend the Florida Constitution to raise the millage cap for the Northwest Florida Water Management District.

9. Regulatory agencies should require a high level of use efficiency as a condition to incentives and long-term permitting assurances.

10. Create incentives for private land owners to capture and store water.

11. Optimize the use of alternative water supplies which take and store peak surface water flows and also restores the natural system.

12. Minimum flows and levels (MFLs) must be set for all surface water bodies where consumptive use permits are sought; natural groundwater levels should not be ignored to the detriment and loss of the dependent natural ecological systems (wetlands and spring systems for example). Natural system ecological needs must not be compromised to meet the public water supply needs. (If MFLs are not achieved, a recovery strategy should be implemented.)

13. Reinstate the annual state funding for alternative water supply development and water quality improvement (i.e., SB 444 funding to be matched by Water Management Districts and local governments/ utilities). Make SB 444 funding a recurring source of annual state funding for alternative water supply development and reinstate original funding levels at a minimum.

14. While protecting water quality, maximize the beneficial use of reclaimed water and improve upon the capture and storage of excess water. Recruit and connect large industrial users to reclaimed water systems to reduce demand on existing and future potable systems. (It is recommended that a goal of 100% beneficial and cost effective reuse of wastewater from publicly owned wastewater treatment facilities be established for the year 2030.) (The management of wastewater needs to continue to evolve from a disposal problem to a valuable water supply opportunity.

15. Support Florida-specific research on climate change and water management interrelationship to better understand the state's water vulnerabilities and make

appropriate and effective adaptations to water planning regulatory and financial programs. This research should include consideration of:

- o Energy and greenhouse emission consequences of water supply activities
- o Increased water use efficiencies to reduce carbon footprints
- o The protection of drinking water and wastewater infrastructure against the threat of rising sea level. FDEP, Water Management Districts, the Florida Energy and Climate Commission, and water managers should fully incorporate climate change implications into their water planning, regulatory, and financial assistance programs and should fully consider the energy and greenhouse gas emissions consequences of water resource and supply activities.

16. Make creation of new water storage (including new reservoirs, ASR, and wet season storage) a statewide priority by prioritizing funding, land acquisition, and needed regulatory reforms (for ASR).
17. Manage stormwater runoff that is discharged into municipal stormwater systems as a valuable water source to be used or reused for conservation purposes such as community irrigation, not as a waste product requiring quick disposal.
18. Encourage Low Impact Development practices, as well as other source control measures to enhance ambient water quality in receiving water bodies.

Source: Century Commission for a Sustainable Florida, Third Annual Report to the Governor and Legislature, January, 2009, 16 pp.

List of Abbreviations and Acronyms

ASR	Aquifer Storage and Recovery
bgd	billion gallons per day
BMPs	Best Management Practices
C&SF Project	Central and Southern Florida Project for Flood Control and Other Purposes
CERP	Comprehensive Everglades Restoration Project
cfs	cubic feet per second
DACS	Department of Agriculture and Consumer Services
DCA	Department of Community Affairs
EAA	Everglades Agricultural Area
EIS	Environmental Impact Statement
EPA	U.S. Environmental Protection Agency
F.A.C.	Florida Administrative Code
FDEP	Florida Department of Environmental Protection
F.S.	Florida Statutes
FWC	Florida Fish and Wildlife Conservation Commission
FY	Fiscal Year
gpcd	gallons per capita per day
mgd	million gallons per day
MFLs	minimum flows and levels
NWFWMD	Northwest Florida Water Management District
ppm	parts per million
ppt	parts per thousand
SFWMD	South Florida Water Management District
SJRWMD	St. Johns River Water Management District
SRWMD	Suwannee River Water Management District
SWFWMD	Southwest Florida Water Management District
STA	Stormwater Treatment Area
TMDL	Total Maximum Daily Load
WMD	water management district
USACE	U.S. Army Corps of Engineers

APPENDIX 4

Glossary [1]

Acre-foot: The volume of water that covers one acre to a depth of one foot; 325,872 gallons.

Ad valorem tax: A tax imposed on the value of real and personal property. Assessed by water management districts and local governments.

Adaptation: Adjustments in human social systems in response to actual or expected climate change.

Adaptive management: The structured application of scientific information and explicit feedback mechanisms to refine and improve future management decisions.

Alternative water supply: A supply of water that has been reclaimed after one or more public supply, municipal, industrial, commercial or agricultural uses; or a supply of stormwater, or brackish or saltwater, that has been treated in accordance with applicable rules and standards sufficient to supply the intended use.

Aquifer: A geologic formation that contains sufficient saturated permeable material to yield significant quantities of water to wells, springs, or surface water.

Aquifer Storage and Recovery (ASR): The injection of fresh water into a confined aquifer during times when excess water exists, and recovering it during times when there is a deficit of water.

Artificial recharge: The intentional addition of water to an aquifer by injection or infiltration (e.g., directing surface water onto spreading basins).

Automatic irrigation system: An irrigation system that operates following a preset program entered into an automatic controller.

Backpumping: The practice of actively pumping water leaving an area back into a surface water body, such as Lake Okeechobee.

Base flow: The sustained low flow of a stream, usually groundwater inflow to the stream channel.

Biofuels: Fuels derived from plant materials that can be regenerated, such as wood, wood waste, agricultural waste, municipal solid waste, and landfill gases. Examples of biofuels are ethanol, methanol, and biodiesel.

Biscayne Aquifer: A portion of the Surficial Aquifer System, which provides most of the fresh water for public water supply and agriculture within Miami-Dade, Broward, and southeastern Palm Beach Counties.

Brackish water: Water with a salinity intermediate between seawater and freshwater.

Carbon dioxide A naturally occurring gas and also a by-product of burning fossil fuels and biomass, as well as land-use changes and other industrial processes. It is the principal anthropogenic greenhouse gas.

Carbon sequestration: The process of removing carbon from the atmosphere and storing it. Trees, for example, absorb carbon dioxide, release oxygen, and store the carbon. Soils can also perform this function under certain circumstances.

Central and Southern Florida Flood Control Project (C&SF Project): A system of canals, storage areas, and water control structures spanning the area from Lake Okeechobee to both the east and west coasts and from Orlando south through the Everglades to Florida Bay. It was designed and constructed primarily during the 1950s and 1960s by the U.S. Army Corps of Engineers to provide flood control and improve navigation and recreation.

Climate: The long-term average weather of a region including typical weather patterns, the frequency and intensity of storms, cold spells, and heat waves. Climate is not the same as weather.

Climate change: Refers to changes in long-term trends in the average climate, such as changes in average temperatures.

Comprehensive Everglades Restoration Plan (CERP): Plan that provides a framework and guide to restore, protect, and preserve the water resources of central and southern Florida, including the Everglades. It covers 16 counties over an 18,000-square-mile area.

Cone of depression: The depression of the water table around a pumping well caused by withdrawal of water.

Conservation rate structure: A schedule of utility water rates designed to promote efficient use of water by providing economic incentives, such as inclining block rates, marginal cost pricing, and seasonal surcharges.

Conserve Florida: The state program for water conservation in public water supply.

Consumptive Use Permit (CUP): A permit issued by a water management district under authority of Chapter 373, F.S., allowing withdrawal of water for consumptive use. (Same as Water Use Permit.)

Declining (decreasing) block rate structure: A pricing structure in which the amount charged per unit of water (e.g., dollars per 1,000 gallons) decreases as customer water consumption increases. This type of rate structure is not considered to be water conserving.

Desalination: A process that treats saline water to remove chlorides and dissolved solids, resulting in the production of fresh water. Primarily used for drinking water.

Domestic withdrawals: Water used for normal household purposes, such as drinking, food preparation, bathing, washing clothes and dishes, flushing toilets, and watering lawns and gardens. The water may be obtained from a public supplier or may be self-supplied. Also called residential water use.

Drainage district: A local district that is created by special act of the legislature and authorized under Ch. 298 F.S., to construct, complete, operate, maintain, and repair any works necessary to implement an adopted water control plan.

Drawdown: The vertical distance a water level is lowered resulting from a withdrawal at a given point.

Drip irrigation: An irrigation system in which water is applied directly to the root zone of plants by means of applicators (emitters, porous tubing, or perforated pipe) operated under low pressure.

Drought rates: Rate structures that impose higher rates during water shortages in order to reduce water use.

Embodied energy: The energy consumed by all the processes associated with the production of a product (such as water) from the acquisition of natural resources to product delivery.

Estuary: A body of water where freshwater and saltwater meet; an arm of the ocean at the lower end of a river.

Eutrophication: The gradual increase in nutrients in a body of water. Natural eutrophication is a gradual process, but human activities may greatly accelerate the process.

Evapotranspiration (ET): The total loss of water to the atmosphere by evaporation from land and water surfaces and by transpiration from plants.

Everglades Forever Act (EFA): A 1994 Florida law (Section 373.4592, F.S.), amended in 2003, to promote Everglades restoration and protection.

Finger-fill canals: Canals created by dredging wetlands; the resulting fill is used to build dry land, usually for houses

First magnitude spring: A spring with a measured flow of at least 100 cubic feet per second.

Flatwoods: Relatively open canopied forests of scattered pine trees and saw palmetto located on flat, poorly drained terrain.

Flood: An overflow or inundation that comes from a river or other body of water and causes or threatens damage.

Flood irrigation: The application of irrigation water whereby the entire surface of the soil is covered by ponded water.

Floodplain: Land area subject to inundation by flood waters from a river, watercourse, lake, or coastal waters. Floodplains are delineated according to their estimated frequency of flooding.

Flood-prone: Land area subject to periodic inundation, whether or not adjacent to a water body.

Florida Administrative Code (F.A.C.): The Florida Administrative Code is the official compilation of the administrative rules and regulations of state agencies.

Florida Department of Environmental Protection (FDEP): The state environmental agency that has general supervisory authority over the water management districts.

Florida Water Plan: The state-level water resource plan developed by the Florida Department of Environmental Protection under Section 373.036, F.S.

Floridan Aquifer System: A highly-used aquifer system composed of the Upper Floridan and Lower Floridan Aquifers, which extends from Florida into Georgia.

Flow: The amount of water flowing by a particular point over some specified time. Flow is frequently expressed in millions of gallons per day (mgd).

Global warming: The progressive gradual rise of the Earth's average surface temperature caused in part by increased concentrations of greenhouse gases in the atmosphere. (Global "climate change" is the more inclusive and accurate term.)

Governing board: Governing body of a Water Management District created under Section 373.069, F.S.

Greenhouse effect: The effect produced as greenhouse gases allow incoming solar radiation to pass through the Earth's atmosphere, but prevent part of the outgoing infrared radiation from the Earth's surface and lower atmosphere from escaping into outer space. If the atmospheric concentrations of these greenhouse gases rise, the average temperature of the lower atmosphere will gradually increase.

Greenhouse gas (GHG): Any gas that contributes to the greenhouse effect. Greenhouse gases include, but are not limited to, water vapor, carbon dioxide (CO_2), methane (CH_4), and nitrous oxide (N_2O).

Groundwater: Water beneath the surface of the ground, whether or not flowing through known and definite channels.

Invasive species: Non-native species of plants or animals that outcompete native species in a specific habitat.

Hurricane: Tropical storm with winds of 74 mph or greater.

Hydrologic cycle: The natural recycling process powered by the sun that causes water to evaporate into the atmosphere, condense, and return to Earth as precipitation.

Hydropattern: The full range of hydrologic parameters, which include the depth of water, duration of inundation, and the timing and distribution of freshwater flow.

Hydroperiod: The frequency and duration of inundation or saturation of an ecosystem.

Intergovernmental Panel on Climate Change (IPCC): The IPCC was established in 1988 by the World Meteorological Organization and the U.N. Environment Program. The IPCC is responsible for providing the scientific and technical foundation for the United Nations Framework Convention on Climate Change (UNFCC), primarily through the publication of periodic assessment reports.

Informative billing: A system of providing water utility customers with useful information on the relationship between the amount of water they use and the cost associated with that use.

Karst: Type of landform modified by dissolution of soluble limestone and characterized by caves, sinkholes, underground drainage systems, and disappearing streams.

Landscape irrigation: Application of water to a landscape by artificial means, that is, means other than natural precipitation.

Lost to tide: The excess amount of water leaving the system that is beyond the amount needed by downstream estuaries.

Microirrigation: The application of small quantities of water directly on or below the soil surface. Microirrigation encompasses a number of methods or concepts including drip, subsurface, microbubbler, and microspray irrigation.

Minimum Flow and Level (MFL): The point at which further withdrawals would cause significant harm to the water resources/ecology of the area.

MILs (Mobile Irrigation Labs) and irrigation evaluations. Evaluations of irrigation systems and practices with advice for improving water use efficiency.

Nonpoint-source water pollution: A diffuse source of pollution, having no single point of origin (such as urban stormwater runoff or leaching of agricultural chemicals from crop land) and which enters the water resource diffusely over a large area.

Potable water: Water that is safe for human consumption.

Prior appropriation: Doctrine of water use common in the western United States whereby the first water user has continued rights to withdraw and use the water.

Public Water Supply: Utilities that provide potable water for public use.

Reasonable-beneficial use: The use of water under Florida law in such quantity as is necessary for economic and efficient utilization for a purpose and in a manner which is both reasonable and consistent with the public interest.

Recharge: Replenishment of groundwater through the infiltration of rainfall and other surface waters.

Recharge area: A place where water is able to seep into the ground and replenish an aquifer because no confining layer is present. Such an area is particularly vulnerable to any pollutants that could be in the water.

Reclaimed water: Water that has received at least secondary treatment and basic disinfection and is reused after flowing out of a domestic wastewater treatment facility.

Regional water supply plan: Detailed water supply plan developed by a water management district under Section 373.0361, F.S., providing an evaluation of available water supply and projected demands, at the regional scale. The planning process projects future demand for 20 years and develops strategies to meet identified needs.

Retrofitting: The replacement of existing water fixtures, appliances, and devices with more efficient fixtures, appliances, and devices for the purpose of conservation.

Reverse osmosis: The process of removing salts from water using a fine membrane. The product water is pushed through a fine membrane by high-pressure pumps. The dissolved salts are unable to pass through. The freshwater is put to use and the remaining salty wastewater must be disposed of.

Riparian rights: A concept of water law under which authorization to use water in a stream is based on ownership of the land adjacent to the stream.

Saltwater intrusion: This occurs when more dense saline water moves laterally inland from the seacoast, or moves vertically upward, to replace fresher water in an aquifer.

Sea level: Long-term average position of the sea surface.

Sequestration (of carbon): The process of removing carbon from the atmosphere and storing it either through biological processes (e.g., plants and trees), or geological processes (through storage of CO_2 in soil or underground reservoirs).

Settlement Agreement: The 1992 court-ordered settlement agreement between the state of Florida and federal parties that directs the clean-up of federal waters within the Everglades, including Everglades National Park and the Loxahatchee National Wildlife Refuge.

Sheetflow: The movement of water, like a sheet, across a surface – moving not in channels, but as a whole mass.

Sink: Any process, activity, or mechanism which removes a greenhouse gas.

Sinkhole: Depression in the land surface caused when rainwater dissolves limestone near the ground surface or when the roofs of underground channels and caverns collapse.

Spring: The natural outflow of water from underground through a break in the land surface caused by the pressure of the groundwater.

Spring-fed river: A type of river with cool, clear water issuing from springs.

Spring group: A collection of individual spring vents and seeps that lie within a discrete spring recharge basin (or springshed).

Spring magnitude: A category based on the volume of flow from a spring per unit time.

Springshed: The total land area that contributes rainfall and runoff to a spring or series of connected springs.

Stormwater runoff: Rainwater that runs off land surfaces into the nearest body of water.

Stormwater Treatment Area (STA): A large, constructed wetland designed to remove pollutants, particularly nutrients such as phosphorus, from stormwater runoff using natural processes.

Submerged aquatic vegetation: Wetland plants that exist completely below the water surface.

Subsidence: The lowering of the soil level caused by the shrinkage of organic layers. This shrinkage is due to a number of factors such as desiccation, consolidation, and biological oxidation.

Surficial aquifer system: Often the principal source of water for urban uses within certain areas of south Florida. This aquifer is unconfined, consisting of varying amounts of limestone and sediments that extend from the land surface downward.

Total Maximum Daily Load (TMDL): The maximum amount of pollution that a water body can assimilate from all sources without violating state water quality standards.

Water Conservation Areas: Three diked areas of the remnant Everglades that are hydrologically controlled for flood control and water supply purposes. Primary targets of the Everglades restoration, and major components of the Everglades Protection Area.

Water management districts: There are five water management districts in Florida created in 1972 by the Florida legislature to administer Chapter 373, F.S.

Water reservations: State law on water reservations, in Section 373.223(4), F.S., defines water reservations as follows: "The governing board or the department, by regulation, may reserve from use by permit applicants, water in such locations and quantities, and for such seasons of the year, as in its judgment may be required for the protection of fish and wildlife, or the public health and safety. Such reservations shall be subject to periodic review and revision in the light of changed conditions. However, all presently existing legal uses of water shall be protected so long as such use is not contrary to the public interest."

Water Use Permit (WUP): See consumptive use permit.

Water or waters in the state: Any and all water on or beneath the surface of the ground or in the atmosphere, including natural or artificial watercourses, lakes, ponds, or diffused surface water and water percolating, standing, or flowing beneath the surface of the ground, as well as all coastal waters within the jurisdiction of the state.

Watershed: The land area that contributes to the flow of water into a receiving body of water, such as a river, lake, or estuary. All lands in the watershed drain toward the river, lake, or estuary.

Wetlands: Ecosystems whose soil is saturated for long periods seasonally or continuously, including marshes and swamps

NOTE

1. Sources of this glossary include the 2002 *Water Conservation Initiative Report* and the glossary prepared for the December 2007 Water Conservation Summit of the South Florida Water Management District; the U.S. Geological Survey *Water Basics Glossary*; and the U.S. EPA *Climate Change Glossary.*

Index